A First Course in Differential Geometry

MONOGRAPHS AND TEXTBOOKS IN PURE AND APPLIED MATHEMATICS

1. *K. Yano,* Integral Formulas in Riemannian Geometry (1970) *(out of print)*
2. *S. Kobayashi,* Hyperbolic Manifolds and Holomorphic Mappings (1970) *(out of print)*
3. *V. S. Vladimirov,* Equations of Mathematical Physics (A. Jeffrey, editor; A. Littlewood, translator) (1970) *(out of print)*
4. *B. N. Pshenichnyi,* Necessary Conditions for an Extremum (L. Neustadt, translation editor; K. Makowski, translator) (1971)
5. *L. Narici, E. Beckenstein, and G. Bachman,* Functional Analysis and Valuation Theory (1971)
6. *D. S. Passman,* Infinite Group Rings (1971)
7. *L. Dornhoff,* Group Representation Theory (in two parts). Part A: Ordinary Representation Theory. Part B: Modular Representation Theory (1971, 1972)
8. *W. Boothby and G. L. Weiss (eds.),* Symmetric Spaces: Short Courses *(out of print)* Presented at Washington University (1972)
9. *Y. Matsushima,* Differentiable Manifolds (E. T. Kobayashi, translator) (1972)
10. *L. E. Ward, Jr.,* Topology: An Outline for a First Course (1972) *(out of print)*
11. *A. Babakhanian,* Cohomological Methods in Group Theory (1972)
12. *R. Gilmer,* Multiplicative Ideal Theory (1972) *(out of print)*
13. *J. Yeh,* Stochastic Processes and the Wiener Integral (1973) *(out of print)*
14. *J. Barros-Neto,* Introduction to the Theory of Distributions (1973) *(out of print)*
15. *R. Larsen,* Functional Analysis: An Introduction (1973) *(out of print)*
16. *K. Yano and S. Ishihara,* Tangent and Cotangent Bundles: Differential Geometry (1973) *(out of print)*
17. *C. Procesi,* Rings with Polynomial Identities (1973)
18. *R. Hermann,* Geometry, Physics, and Systems (1973)
19. *N. R. Wallach,* Harmonic Analysis on Homogeneous Spaces (1973) *(out of print)*
20. *J. Dieudonné,* Introduction to the Theory of Formal Groups (1973)
21. *I. Vaisman,* Cohomology and Differential Forms (1973)
22. *B.-Y. Chen,* Geometry of Submanifolds (1973) *(out of print)*
23. *M. Marcus,* Finite Dimensional Multilinear Algebra (in two parts) (1973, 1975)
24. *R. Larsen,* Banach Algebras: An Introduction (1973)
25. *R. O. Kujala and A. L. Vitter (eds.),* Value Distribution Theory: Part A; Part B. Deficit and Bezout Estimates by Wilhelm Stoll (1973)
26. *K. B. Stolarsky,* Algebraic Numbers and Diophantine Approximation (1974)
27. *A. R. Magid,* The Separable Galois Theory of Commutative Rings (1974)
28. *B. R. McDonald,* Finite Rings with Identity (1974)
29. *J. Satake,* Linear Algebra (S. Koh, T. Akiba, and S. Ihara, translators) (1975)

30. *J. S. Golan,* Localization of Noncommutative Rings (1975)
31. *G. Klambauer,* Mathematical Analysis (1975)
32. *M. K. Agoston,* Algebraic Topology: A First Course (1976)
33. *K. R. Goodearl,* Ring Theory: Nonsingular Rings and Modules (1976)
34. *L. E. Mansfield,* Linear Algebra with Geometric Applications: Selected Topics (1976)
35. *N. J. Pullman,* Matrix Theory and Its Applications (1976)
36. *B. R. McDonald,* Geometric Algebra Over Local Rings (1976)
37. *C. W. Groetsch,* Generalized Inverses of Linear Operators: Representation and Approximation (1977)
38. *J. E. Kuczkowski and J. L. Gersting,* Abstract Algebra: A First Look (1977)
39. *C. O. Christenson and W. L. Voxman,* Aspects of Topology (1977)
40. *M. Nagata,* Field Theory (1977)
41. *R. L. Long,* Algebraic Number Theory (1977)
42. *W. F. Pfeffer,* Integrals and Measures (1977)
43. *R. L. Wheeden and A. Zygmund,* Measure and Integral: An Introduction to Real Analysis (1977)
44. *J. H. Curtiss,* Introduction to Functions of a Complex Variable (1978)
45. *K. Hrbacek and T. Jech,* Introduction to Set Theory (1978) *(out of print)*
46. *W. S. Massey,* Homology and Cohomology Theory (1978)
47. *M. Marcus,* Introduction to Modern Algebra (1978)
48. *E. C. Young,* Vector and Tensor Analysis (1978)
49. *S. B. Nadler, Jr.,* Hyperspaces of Sets (1978)
50. *S. K. Sehgal,* Topics in Group Rings (1978)
51. *A. C. M. van Rooij,* Non-Archimedean Functional Analysis (1978)
52. *L. Corwin and R. Szczarba,* Calculus in Vector Spaces (1979)
53. *C. Sadosky,* Interpolation of Operators and Singular Integrals: An Introduction to Harmonic Analysis (1979)
54. *J. Cronin,* Differential Equations: Introduction and Quantitative Theory (1980)
55. *C. W. Groetsch,* Elements of Applicable Functional Analysis (1980)
56. *I. Vaisman,* Foundations of Three-Dimensional Euclidean Geometry (1980)
57. *H. I. Freedman,* Deterministic Mathematical Models in Population Ecology (1980)
58. *S. B. Chae,* Lebesgue Integration (1980)
59. *C. S. Rees, S. M. Shah, and Č. V. Stanojević,* Theory and Applications of Fourier Analysis (1981)
60. *L. Nachbin,* Introduction to Functional Analysis: Banach Spaces and Differential Calculus (R. M. Aron, translator) (1981)
61. *G. Orzech and M. Orzech,* Plane Algebraic Curves: An Introduction Via Valuations (1981)
62. *R. Johnsonbaugh and W. E. Pfaffenberger,* Foundations of Mathematical Analysis (1981)

63. *W. L. Voxman and R. H. Goetschel,* Advanced Calculus: An Introduction to Modern Analysis (1981)
64. *L. J. Corwin and R. H. Szczarba,* Multivariable Calculus (1982)
65. *V. I. Istrăţescu,* Introduction to Linear Operator Theory (1981)
66. *R. D. Järvinen,* Finite and Infinite Dimensional Linear Spaces: A Comparative Study in Algebraic and Analytic Settings (1981)
67. *J. K. Beem and P. E. Ehrlich,* Global Lorentzian Geometry (1981)
68. *D. L. Armacost,* The Structure of Locally Compact Abelian Groups (1981)
69. *J. W. Brewer and M. K. Smith, eds.,* Emmy Noether: A Tribute to Her Life and Work (1981)
70. *K. H. Kim,* Boolean Matrix Theory and Applications (1982)
71. *T. W. Wieting,* The Mathematical Theory of Chromatic Plane Ornaments (1982)
72. *D. B. Gauld,* Differential Topology: An Introduction (1982)
73. *R. L. Faber,* Foundations of Euclidean and Non-Euclidean Geometry (1983)
74. *M. Carmeli,* Statistical Theory and Random Matrices (1983)
75. *J. H. Carruth, J. A. Hildebrant, and R. J. Koch,* The Theory of Topological Semigroups (1983)
76. *R. L. Faber,* Differential Geometry and Relativity Theory: An Introduction (1983)
77. *S. Barnett,* Polynomials and Linear Control Systems (1983) 1983)
78. *G. Karpilovsky,* Commutative Group Algebras (1983)
79. *F. Van Oystaeyen and A. Verschoren,* Relative Invariants of Rings: The Commutative Theory (1983)
80. *I. Vaisman*, A First Course in Differential Geometry (1984)

Other Volumes in Preparation

A First Course in Differential Geometry

Izu Vaisman
University of Haifa
Haifa, Israel

CRC Press
Taylor & Francis Group
Boca Raton London New York

CRC Press is an imprint of the
Taylor & Francis Group, an **informa** business

First published 1984 by Marcel Dekkar, Inc.

Published 2019 by CRC Press
Taylor & Francis Group
6000 Broken Sound Parkway NW, Suite 300
Boca Raton, FL 33487-2742

First issued in paperback 2019

ISBN 13: 978-0-367-45187-5 (pbk)
ISBN 13: 978-0-8247-7063-1 (hbk)

Visit the Taylor & Francis Web site at
http://www.taylorandfrancis.com

and the CRC Press Web site at
http://www.crcpress.com

Library of Congress Cataloging in Publication Data

Vaisman, Izu.
A first course in differential geometry.

(Pure and applied mathematics ; 80)
Includes bibliographical references and index.
1. Geometry, Differential. I. Title. II. Series:
Pure and applied mathematics (Marcel Dekker, Inc.) ; 80.
QA641.V26 1984 516.3'6 83-15306
ISBN 0-8247-7063-3

Preface

Many excellent textbooks on differential geometry, at various levels, are available. However, the instructor may still have difficulty in choosing one suitable for a particular audience, since differential geometry is taught under many different circumstances.

Therefore, we propose a new approach which is designed to serve as an introductory course in differential geometry for advanced undergraduate students. This book is based on lectures given by the author at several universities. The distinguishing feature of the text is its more ample involvement of differentiable manifolds, which led to a specific choice of the topics included.

The text is divided into three chapters: Differentiable Manifolds in $\mathbb{R}^n$, Curves in E^2 and E^3, and Surfaces in E^3; each chapter is in turn divided into several sections. Each section is informally subdivided into parts, which, we hope, will facilitate an understanding of the text. Each section is followed by a number of exercises, whose aim is to permit the reader to interact actively with the text. The exercises are not original, and most of them can be found in any book on differential geometry. In particular, they can be found in the citations in the References.

Because the topics presented are classic, we did not provide bibliographical references except for a few isolated instances. This explains the shortness of the list of references at the end of the book. However, among the citations we have included some texts in calculus, topology, and linear algebra, primarily to be used as a refresher material.

Izu Vaisman

Contents

A First Course in Differential Geometry

1

Differentiable Manifolds in $\mathbb{R}^n$

1.1 THE SPACE $\mathbb{R}^n$

Differential geometry studies geometric figures and spaces that can be characterized by means of differentiable functions. Consequently, the basic reference space is $\mathbb{R}^n$. The space $\mathbb{R}^n$ has many important structures and we recall some of them here.

The n-dimensional real arithmetic space $\mathbb{R}^n$ ($n \neq 0$) is the set

$$\mathbb{R}^n = \underbrace{\mathbb{R} \times \cdots \times \mathbb{R}}_{n \text{ times}} = \{(x^1, \ldots, x^n) \mid x^i \in \mathbb{R};\ i = 1, \ldots, n\} \tag{1.1}$$

where $\mathbb{R}$ denotes the real field, n is a fixed natural number, and the product in (1.1) is the cartesian product of sets. The elements of $\mathbb{R}^n$ are called points and the corresponding numbers x^i are the coordinates. For $n = 0$ one puts $\mathbb{R}^0 = \{0\}$.

Thus the arithmetic space $\mathbb{R}^n$ is defined as a set. However, we may attach to this space many interesting structures. Let us make the general convention of denoting the points of $\mathbb{R}^n$ by the letters x, y, ..., and their coordinates by the corresponding indexed letters (x^i), (y^i), ... $(i = 1, \ldots, n)$.

A first important structure is obtained if we define the distance between two points by

$$\rho(x, y) = \left[\sum_{i=1}^{n} (y^i - x^i)^2\right]^{\frac{1}{2}} \tag{1.2}$$

a formula that actually goes back to Pythagoras's theorem. This distance makes $\mathbb{R}^n$ into a metric space, and the pair $(\mathbb{R}^n, \rho)$ will be called the metric space $\mathbb{R}^n$. Implicitly, $\mathbb{R}^n$ gets a topological structure, which is characterized by the fact that, for every point x, the open balls

$$B(x,\epsilon) = \{y \in \mathbb{R}^n \mid \rho(x,y) < \epsilon\} \tag{1.3}$$

define a basis of open neighborhoods.

We should also mention the important subsets of the metric space $\mathbb{R}^n$, defined by

$$S^{n-1}(x,\ell) = \left\{ y \in \mathbb{R}^n \,\middle|\, \sum_{i=1}^{n} (y^i - x^i)^2 = \ell^2 \right\} \tag{1.4}$$

$S^{n-1}(x,\ell)$ is called the (n - 1) sphere with center x and radius ℓ. In particular, $S^{n-1}((0,\ldots,0),\ 1)$ is the sphere of radius 1 and center O = (0, ..., 0)—a point called the origin of $\mathbb{R}^n$; it will be denoted simply by S^{n-1} and will be called the unit (n - 1) sphere.

Second, we can give the arithmetic space $\mathbb{R}^n$ a natural structure of an n-dimensional linear space over the real field $\mathbb{R}$ by defining

$$\begin{aligned} (x^i) + (y^i) &= (x^i + y^i) \\ \lambda(x^i) &= (\lambda x^i) \qquad \lambda \in \mathbb{R} \end{aligned} \tag{1.5}$$

In this case we shall call the elements of $\mathbb{R}^n$ vectors, and introduce the scalar product

$$xy = x \cdot y = \sum_{i=1}^{n} x^i y^i \tag{1.6}$$

which yields a norm (length) of vectors

$$|x| = \left[\sum_{i=1}^{n} (x^i)^2 \right]^{\frac{1}{2}} \tag{1.7}$$

and an angle between two vectors defined by

$$\cos \varphi = \frac{x \cdot y}{|x|\ |y|} \tag{1.8}$$

All the classical notions of linear algebra, such as linear dependence, bases and coordinates, linear subspaces, and so on, may then be used.

Next, in order to point out a third basic structure, let us temporarily denote by $\mathbb{R}p^n$ the metric space $\mathbb{R}^n$ and by $\mathbb{R}\ell^n$ the linear space $\mathbb{R}^n$.* Then we can define a map

$$\pi : \mathbb{R}p^n \times \mathbb{R}p^n \to \mathbb{R}\ell^n \tag{1.9}$$

---

*We use such a notation wherever the common notation $\mathbb{R}^n$ might be confusing.

by means of the formula

$$\pi(x,y) = (y^i - x^i) \tag{1.10}$$

This mapping has the following two properties:

(a) $\pi(x,y) + \pi(y,z) = \pi(x,z)$.

(b) $\pi | \{0\} \times \mathbb{R}p^n$ is a bijection.

We shall denote $\pi(x,y) = \underline{xy}$.

Correspondingly, we say that $\mathbb{R}^n$ has a structure of an n-dimensional euclidean space consisting of the system $(\mathbb{R}p^n, \mathbb{R}\ell^n, \pi, \cdot)$. $\mathbb{R}^n$ together with this structure will be called the euclidean space $\mathbb{R}^n$.

More generally, a system $(E^n, V^n, \pi, \cdot)$, where E^n is a set, V^n is a real linear n-dimensional vector space endowed with the scalar product $\cdot$, and $\pi: E^n \times E^n \to V^n$ is a map satisfying the properties (a) and (b) with respect to some point $O \in E^n$, is called an n-dimensional euclidean space E^n. The elements of E^n are then called points and those of V^n vectors. Hereafter, the vectors of euclidean spaces will be denoted by underscores. This definition is equivalent to the usual axiomatic geometrical definition of euclidean space, and the geometric viewpoint requires the study of the objects lying in such spaces.

In a euclidean space E^n, one defines a (cartesian) frame by an origin O, which is a point of E^n, and an orthonormal basis $\{\underline{e}_i\}$ $(i = 1, \ldots, n)$ of V^n. Then every point $p \in E^n$ has the radius vector $\pi(O,p)$ and the coordinates of p are the coordinates of $\pi(O,p)$ with respect to the chosen basis. Hence the choice of a frame provides, via coordinates, an isomorphism of E^n with the euclidean space $\mathbb{R}^n$, and the last is considered to be representative of all the E^n.

However, the isomorphism constructed above is not unique. Indeed, if we take another frame $\{O', \underline{e}'_i\}$, we have $\underline{e}_j = \Sigma_{i=1}^n a^i_j \underline{e}'_i$, where (a^i_j) is an orthogonal matrix. On the other hand, we have

$$\pi(O',p) = \pi(O',O) + \pi(O,p)$$

for $p \in E^n$, and, if p has coordinates (x^i) and (x'^i) with respect to the two frames, respectively, this yields

$$x'^i = \sum_{j=1}^n a^i_j x^j + b^i \tag{1.11}$$

where $\pi(O',O) = \Sigma_{i=1}^n b^i \underline{e}'_i$. We say that the euclidean properties of E^n are those properties of $\mathbb{R}^n$ which are invariant under the transformation (1.11).

For instance, we can define an h plane in E^n as a subset characterized by the equations

$$\sum_{i=1}^{n} a_i^\sigma x^i + b^\sigma = 0 \qquad \sigma = 1, \ldots, n-h \tag{1.12}$$

where rank $(a_i^\sigma) = n - h$, and this is a geometric notion because the form of the system (1.12) is unchanged by (1.11). It is to be remarked here that the euclidean structure of E^n induces a euclidean structure in any h plane, and this structure is associated with an h-dimensional linear subspace of the vector space of E^n. For $h = 1, 2, n-1$, we have a line, a plane, and a hyperplane, respectively. Other euclidean figures are the spheres defined by equations of the form appearing in formula (1.4).

Finally, let us mention that the point transformations, defined in E^n by equations of the form (1.11), where (x^i) and (x'^j) are now the coordinates of a point and of its image, also have a euclidean character. These are called orthogonal transformations. Since the orthogonal transformations and the euclidean coordinate transformations are expressed by the same formula (1.11), one can also say that euclidean geometry consists of the properties of euclidean space $\mathbb{R}^n$ which are invariant under orthogonal transformations.

In particular, for $n = 2, 3$ one gets the usual euclidean plane and space geometry, assumed to be familiar to the reader. In these cases the vectors are the usual geometric vectors, and the usual operations, such as scalar product, vector product, and mixed product of vectors, have a geometric character.

EXERCISES

1.1 Prove that the function ρ defined by formula (1.2) satisfies the properties of a metric:

1. $\rho(x,y) \geq 0$, equality holding iff $x = y$
2. $\rho(x,y) = \rho(y,x)$ (symmetry)
3. $\rho(x,z) \leq \rho(x,y) + \rho(y,z)$ (triangle inequality)

1.2 Prove that $\cos\varphi$ defined by formula (1.8) satisfies the condition $|\cos\varphi| \leq 1$.

1.3 Let us denote

$$E^h = \left\{ v \mid v \in \mathbb{R}^n,\ v = v_0 + \sum_{\alpha=1}^{h} t^\alpha v_\alpha \right\}$$

$$V^h = \{ w \mid w = u_2 - u_1,\ u_1, u_2 \in E^h \}$$

where v_0, $v_\alpha \in \mathbb{R}^n$, and t^α are arbitrary real numbers. Define $\pi : E^h \times E^h \to V^h$ by $\pi(u_1, u_2) = u_2 - u_1$, and use the classical scalar product $\cdot$ in $\mathbb{R}^n$. Prove that the system $(E^h, V^h, \pi, \cdot)$ is an h-dimensional euclidean space.

1.4 Let $(E^n, V^n, \pi, \cdot)$ be a euclidean space, and let H be an h plane of E^n. Define in detail the induced euclidean structure of H.

1.5 A subset Σ of a euclidean space E^n is called an m-sphere if it is the intersection of an (m + 1)-plane with a sphere of E^n. Write the general form of the equations of an m-sphere with respect to a frame. Prove that the form of these equations remains unchanged by the coordinate transformations (1.11)

1.2 DIFFERENTIABLE FUNCTIONS

For now, we shall fix our attention on the topological metric space $\mathbb{R}^n$. The reader should be familiar with the classical properties of differentiable functions [e.g., Kaplan (1968)]. In any case, some of them will be reviewed here in an appropriate form.

DEFINITION 1.1 A function $f : M \to \mathbb{R}$, defined on some subset $M \subset \mathbb{R}^n$, is said to be differentiable if there is some open subset $U \subseteq \mathbb{R}^n$ and some function $F : U \to \mathbb{R}$ such that

(a) $M \subseteq U$ and $f = F | M$.
(b) F has finite continuous partial derivatives of any order h = 1, 2, ... at all the points of U.

Here, the necessity of extending f to a function F defined on an open set U follows from the need to assign a proper definition to the partial derivatives. These derivatives appear since F has the form $y = F(x^1, \ldots, x^n)$, where the x^i are the coordinates in $\mathbb{R}^n$ and y is the coordinate on $\mathbb{R}$.

To be more precise, a function f that is differentiable in the sense of Definition 1.1 is said to be of class C^∞. Similarly, if we replace condition (b) of Definition 1.1 by the demand that F have continuous derivatives up to the order k (k = 1, 2, ...), we get the notion of a function of class C^k. A function f that is only continuous is said to be of class C^0. A function f for which F is real analytic (i.e., admits a Taylor series development at every $x \in U$) is said to be of class C^ω. For the sake of simplicity, we have chosen to consider only C^∞ differentiability, in spite of the fact that most of the results will be valid for the class C^k, where k is sufficiently large. Therefore, in the absence of specific differentiability assumptions, differentiable will mean of class C^∞, in agreement with Definition 1.1.

Now, consider a function $f: M \to \mathbb{R}^k$, where $M \subseteq \mathbb{R}^n$. If x^i $(i = 1, \ldots, n)$ and y^α $(\alpha = 1, \ldots, k)$ are the coordinates in $\mathbb{R}^n$ and $\mathbb{R}^k$, respectively, f can be expressed by the system

$$y^\alpha = y^\alpha(x^i) \tag{1.13}$$

where, in general, each y^α is a function of all the x^i (and this notation convention will always be used in the sequel).

DEFINITION 1.2 The function $f: M \to \mathbb{R}^k$ above is differentiable (or of some class C^k) if all the functions $y^\alpha : M \to \mathbb{R}$ of (1.13) are differentiable.

For instance, it is easy to see that the identity maps and the inclusion maps of subsets into $\mathbb{R}^k$ are differentiable and that a composition of differentiable functions is differentiable as well (if it exists).

Now, consider a differentiable map $f: M \to \mathbb{R}^k$ $(M \subseteq \mathbb{R}^n)$, which can be represented by (1.13) and let x be an interior point of M. Let $\underline{\xi} \in \mathbb{R}^n$ be an arbitrary vector of $\mathbb{R}^n$. Because of the linear structures on $\mathbb{R}^n$ and $\mathbb{R}^k$, the following expression makes sense:

$$\lim_{t \to 0} \frac{f(\underline{x} + t\underline{\xi}) - f(\underline{x})}{t} \tag{1.14}$$

where, of course, the limit of a vector has to be taken componentwise.

Computing componentwise and using the mean value theorem [Fleming (1968)], it follows that the limit (1.14) exists, and it is a vector $\underline{\eta} \in \mathbb{R}^k$ whose coordinates are given by

$$\eta^\alpha = \sum_{i=1}^{n} \frac{\partial y^\alpha}{\partial x^i} \xi^i \tag{1.15}$$

ξ^i here representing the coordinates of $\underline{\xi}$. We thus get a linear map

$$f'_x : \mathbb{R}^n \to \mathbb{R}^k \tag{1.16}$$

whose matrix is the jacobian matrix $(\partial y^\alpha / \partial x^i)$.

DEFINITION 1.3 The linear map f'_x defined above is called the differential or derivative of the differentiable map f at the point $x \in \text{int } M$. The rank of this map is called the rank of f at x. The point x is called nonsingular if $\text{rank}_x f = \min(n, k)$ and singular if $\text{rank}_x f < \min(n, k)$.

Now the known properties and utilizations of the jacobian matrix can be reformulated in terms of the derivative mapping. First, we have

PROPOSITION 1.1
(a) The derivative of an inclusion is the identical linear map.
(b) The derivative of a composite map is given by

$$(g \circ f)'_x = g'_{f(x)} \circ f'_x.$$

Proof: Part (a) follows from the fact that the equations of an inclusion are of the form $y^i = x^i$. Part (b) is a straightforward consequence of the chain rule for partial derivatives.

The jacobian matrix enters essentially into the <u>implicit function theorem</u>. Suppose that $f: U \to \mathbb{R}^k$ is a differentiable mapping from an open subset $U \subseteq \mathbb{R}^n$ to $\mathbb{R}^k$, where $n \geq k$. By a <u>decomposition</u> $\mathbb{R}^n = \mathbb{R}^k \times \mathbb{R}^{n-k}$ we shall understand any bijection of the two sets obtained by some grouping of the coordinates of the points of $\mathbb{R}^n$. For example,

$$(x^1, \ldots, x^n) \longmapsto ((x^{i_1}, \ldots, x^{i_k}), (x^{i_{k+1}}, \ldots, x^{i_n}))$$

where $(i_1, \ldots, i_n)$ is a permutation of $(1, \ldots, n)$, defines such a decomposition. Then the following basic theorem is known to hold:

THEOREM 1.1 (Implicit Function Theorem) Let $f: U \to \mathbb{R}^k$ be a differentiable function as above and let x_0 be a nonsingular point of f. Then there is some decomposition $\mathbb{R}^n = \mathbb{R}^k \times \mathbb{R}^{n-k}$ for which $x_0 = (x_0', x_0'')$, there are some open neighborhoods A of x_0' in $\mathbb{R}^k$, B of x_0'' in $\mathbb{R}^{n-k}$ such that $A \times B \subseteq U$, and there is a unique function $\varphi: B \to A$ which satisfies the conditions
(a) $x_0' = \varphi(x_0'')$
(b) $f(\varphi(y), y) = 0$ for every $y \in B$
Moreover, φ is a differentiable function.

The reader is strongly advised to compare this version of the implicit function theorem with the usual versions in calculus books [e.g., Kaplan [1968)]. Note that the condition of x_0 being nonsingular is equivalent to the existence of a nonvanishing minor of order k in the jacobian matrix of the function f.

As a consequence of this theorem one gets

THEOREM 1.2 (Inverse Function Theorem) Let $f: U \to \mathbb{R}^n$ be a differentiable function defined on the open subset $U \subseteq \mathbb{R}^n$ and let $x_0 \in U$ be a nonsingular point of f. Then there is an open neighborhood $U' \subseteq U$ of x_0, an open neighborhood $V \subseteq f(U)$ of $f(x_0)$, and a unique function $g: V \to U'$ such that $g = (f \mid U')^{-1}$. Moreover, the function g is differentiable.

Indeed, let us define

$$\bar{f}: U \times \mathbb{R}^n \to \mathbb{R}^n$$

by $\bar{f}(x,y) = y - f(x)$, $x \in U$, $y \in \mathbb{R}^n$. Then our hypotheses imply that $(x_0, f(x_0))$ is a nonsingular point of $\bar{f}$ and that (x,y) is a decomposition, as given by Theorem 1.1. Since $\bar{f}(x_0, f(x_0)) = 0$, Theorem 1.2 follows by applying Theorem 1.1 for $\bar{f}$ at the point $(x_0, f(x_0))$.

The inverse function theorem implies

COROLLARY 1.1 Let $f: U \to \mathbb{R}^n$ be a differentiable one-to-one mapping defined on the open subset $U \subseteq \mathbb{R}^n$, for which all the points of U are nonsingular. Then f(U) is open in $\mathbb{R}^n$ and $f^{-1}: f(U) \to U$ is a differentiable function. Moreover, $(f^{-1})'_{f(x)} = (f'_x)^{-1}$.

Proof: The first part of this corollary follows from Theorem 1.2 since, by the uniqueness part of that theorem, f^{-1} coincides with the local differentiable inverses g of f. The second part follows clearly from Proposition 1.1, and it is left as an exercise for the reader.

DEFINITION 1.4 A function $f: U \to V$, where U and V are open subsets of $\mathbb{R}^n$, is called a diffeomorphism if f is a bijection and both f and f^{-1} are differentiable.

It follows, by Corollary 1.1, that if $f: U \to \mathbb{R}^n$, where U is an open subset of $\mathbb{R}^n$, is differentiable and without singular points (i.e., rank $f = n$ everywhere), f is a diffeomorphism of U onto f(U). Conversely, if f is a diffeomorphism, f^{-1} exists and it is differentiable. From $f^{-1} \circ f = \mathrm{id}$ it follows by Proposition 1.1 that $(f^{-1})'_{f(x)} \circ f'_x = \mathrm{id}$, whence f'_x must be of rank n.

These remarks lead to the following useful extension of Definition 1.4.

DEFINITION 1.5 A function $f: U \to \mathbb{R}^k$, where U is an open subset of $\mathbb{R}^n$ with $n \leq k$, is called a diffeomorphism of U onto f(U) (which, generally, is not open in $\mathbb{R}^k$) if f is without singular points and it maps U homeomorphically onto f(U).

Let us explain in more detail the meaning of the last condition of Definition 1.5. In topology, a function $f: U \to f(U)$ is said to be a homeomorphism if it is a bijection, and if both f and f^{-1} are continuous. In our case, the meaning of the continuity of f is clear since U is an open subset of $\mathbb{R}^n$. The continuity of f^{-1} will be defined in the usual topological way, but with the convention that the open neighborhoods of f(U) are the intersection of f(U) with the open neighborhoods of $\mathbb{R}^k$. In other words, f(U) is endowed with the topology induced by $\mathbb{R}^k$. Note that a similar condition was not explicitly formulated in Definition 1.4, since there we could rely on the inverse function theorem.

There is also a localization of the preceding notion: $f: U \to \mathbb{R}^k$ above is called a local diffeomorphism if every point $x \in U$ has an open neighborhood $V \subseteq U$ such that $f|V$ is a diffeomorphism. For example, it follows by Theorem 1.2 that, in the case $k = n$, f is a local diffeomorphism iff $\mathrm{rank}_x f = n$ for every point $x \in U$. Examples of diffeomorphisms and local diffeomorphisms will arise in the exercises and in the following two sections.

We close this section with two additional formal remarks. First, the notion of differentiability and related properties can be extended to functions between euclidean spaces. Indeed, by using coordinates, such functions can be expressed by means of functions between spaces $\mathbb{R}^n$, and the differentiability properties of these last functions are preserved by coordinate transformations of the form (1.11), which are themselves differentiable. The reader is asked to provide details.

Second, we represented a function $f: U \to \mathbb{R}^k$ by the relations (1.13). But often, it will be convenient to write directly the vector

$$\underline{y} = \underline{f}(x^i) \tag{1.17}$$

whose coordinates are the y^α of (1.13). In that case we shall have to use such classical notions as limits, continuity, and differentiability for vector functions of scalar arguments. The general principle is that we use these notions coordinatewise.

There is no difficulty in transposing the usual calculus theorems to such vector functions. For instance, by componentwise application of the Taylor formula, we get the Taylor formula for (1.17):

$$\underline{f}(x^i) = \underline{f}(x_0^i) + \sum_{h=1}^{m} \frac{1}{h!}\left[\sum_{i=1}^{n} \underline{f}_{x_0^i}(x^i - x_0^i)\right]^{(h)} + \underline{\epsilon} \tag{1.18}$$

where $\underline{f}_{x_0^i} = (\partial \underline{f}/\partial x^i)_0$, (h) is the usual symbolic exponent of the Taylor formula in several variables, and $\underline{\epsilon}$ is a vector whose coordinates are of the order $o((x^i - x_0^i)^m)$.

If such a calculus is used in the euclidean plane E^2 or in the space E^3, then the derivatives of the scalar and vector products can be calculated by the Leibnitz rule. The same is true for scalar products in any E^n. Again, the reader is asked to provide details.

EXERCISES

1.6 Consider the function $f: \mathbb{R} \to \mathbb{R}$ defined by

$$f(t) = \begin{cases} 0 & \text{for } t \leq 0 \\ e^{-1/t} & \text{for } t > 0 \end{cases}$$

Prove that this function is everywhere of class C^∞, but that it is not C^ω at $t = 0$. [Hint: The second assertion follows from the fact that $f^{(n)}(0) = 0$ for every n.]

1.7 Define $\varphi: \mathbb{R}^2 \to \mathbb{R}^3$ by

$$x^1 = u^1 \cos u^2 \qquad x^2 = u^1 \sin u^2 \qquad x^3 = ku^1 \qquad k = \text{const}$$

Write the equations of φ_0', and compute its rank.

1.8 Define $\psi: \mathbb{R}^2 \to \mathbb{R}^3$ by the following procedure:

1. Identify $\mathbb{R}^2$ with the subspace of $\mathbb{R}^3$ defined by $x^3 = 0$.
2. Put $\psi(u) = S^2 \cap \overline{nu}$, where S^2 is the unit sphere with center i, $n = (0,0,1)$, and $\overline{nu}$ is the straight line joining the point n to $u \in \mathbb{R}^2$.

Write the equations of ψ and the equations of ψ_0'.

1.9 Prove that the function ψ of Exercise 1.8 is a diffeomorphism of $\mathbb{R}^2$ onto its image.

1.10 Consider the function $f: \mathbb{R} \to \mathbb{R}^2$ defined by

$$x^1 = \cos t \qquad x^2 = \sin t$$

Prove that this function is a local diffeomorphism, but it is not a diffeomorphism.

1.11 Let $\varphi: E^m \to E^n$ be a differentiable function between two euclidean spaces. Define its differential $\varphi_x': V^m \to V^n$, and prove that it is invariant. ($x \in E^m$; V^m and V^n are the vector spaces of the two euclidean spaces.) [Hint: Take two coordinate expressions of φ, $y^\alpha = \varphi^\alpha(x^i)$ and $\tilde{y}^\alpha = \tilde{\varphi}^\alpha(\tilde{x}^i)$, and use the orthogonal coordinate transformations between x^i, $\tilde{x}^i$ and y^α, $\tilde{y}^\alpha$, respectively.]

1.12 Let $\underline{u} = \underline{u}(t)$, $\underline{v} = \underline{v}(t)$, and $\underline{w} = \underline{w}(t)$ be vector functions from $\mathbb{R}$ to either $\mathbb{R}^3$ or E^3. Prove the formulas

$$(\underline{v} \cdot \underline{w})' = \underline{v}' \cdot \underline{w} + \underline{v} \cdot \underline{w}'$$

$$(\underline{v} \times \underline{w})' = \underline{v}' \times \underline{w} + \underline{v} \times \underline{w}'$$

$$(\underline{u}, \underline{v}, \underline{w})' = (\underline{u}', \underline{v}, \underline{w}) + (\underline{u}, \underline{v}', \underline{w}) + (\underline{u}, \underline{v}, \underline{w}')$$

where the prime denotes d/dt, $\cdot$ is the scalar product, $\times$ is the vector (cross) product, and $(\cdot, \cdot, \cdot)$ is the mixed product of vectors.

1.13 Consider a function $\underline{v}: \mathbb{R} \to E^3$. Prove that if $\underline{v} \times \underline{v}' \equiv 0$, then the vector $\underline{v}$ is parallel to a fixed line, and if $(\underline{v}, \underline{v}', \underline{v}'') \equiv 0$, the vector $\underline{v}$ is parallel to a fixed plane. [Hint: To prove the second assertion, use the first one for the vector $\underline{v} \times \underline{v}'$.]

1.3 DIFFERENTIABLE MANIFOLDS IN $\mathbb{R}^n$

The differentiable manifolds in $\mathbb{R}^n$ are subsets, which generalize the intuitively known curves and surfaces. To obtain corresponding m-dimensional objects, $0 \leq m \leq n$, suitable for the methods of differential geometry, we use two principles:

1. The points of such objects must have only m independent coordinates, which can be obtained by giving n - m independent relations between the n coordinates in $\mathbb{R}^n$.
2. The relations mentioned in principle 1 have to be given by differentiable functions.

Thus we are led to

DEFINITION 1.6 A nonempty subset Γ of $\mathbb{R}^n$ is called an elementary m-dimensional subset ($m \geq 0$) if $\Gamma \subseteq U$, where U is an open subset of $\mathbb{R}^n$ and if there are n - m differentiable functions $F_\sigma : U \to \mathbb{R}$ ($\sigma = 1, \ldots, n - m$) such that

(a) $\Gamma = \{x \in U \mid F_\sigma(x) = 0\}$

(b) $\operatorname{rank}(\partial F_\sigma / \partial x^i) = n - m \quad (i = 1, \ldots, n;\ \sigma = 1, \ldots, n - m)$

In other words, there is a differentiable map $F : U \to \mathbb{R}^{n-m}$ which has no singular points and satisfies condition 1. Condition 2 expresses the independence of the equations of Γ. $F_\sigma(x) = 0$ are called the implicit equations of Γ.

DEFINITION 1.7 A nonempty subset $M \subseteq \mathbb{R}^n$ is called an m-dimensional differentiable manifold (or submanifold) embedded in $\mathbb{R}^n$ ($m \geq 0$) if every point $x \in M$ has an open neighborhood U in $\mathbb{R}^n$ for which $U \cap M$ is an elementary m-dimensional subset. The empty set $\emptyset$ is taken as a (-1)-dimensional differentiable manifold.

DEFINITION 1.8 A nonempty subset $N \subseteq \mathbb{R}^n$ is called an m-dimensional manifold (or submanifold) immersed in $\mathbb{R}^n$ ($m \geq 0$) if every point $x \in N$ has an open neighborhood U in $\mathbb{R}^n$ such that $U \cap N = \cup \Gamma_\alpha$, where Γ_α is an at most countable family of m-dimensional elementary subsets of $\mathbb{R}^n$.

Equivalently, N is an immersed manifold iff it is a countable union of elementary subsets. Indeed, if this happens, we can take $U = \mathbb{R}^n$ in the definition above. Conversely, if N satisfies the definition, since $\mathbb{R}^n$ has a countable basis (consisting of spheres with rational radii and whose centers have rational coordinates), we can write N as a countable union of elementary subsets.

Following Sec. 1.2, we might also say that our manifolds are of class C^∞ and, in a similar manner, we might define C^k manifolds for any $k = 1, \ldots, \infty, \omega$; in the last case (i.e., C^ω) these are called real analytic manifolds. Later we shall make some remarks about the possibility of defining C^0 manifolds. In this book all manifolds will be C^∞ unless otherwise specified.

Now we shall provide some examples.

(a) $m = 0$. Theorem 1.2 shows that a zero-dimensional elementary subset Γ is discrete; that is, every $x \in \Gamma$ has an open neighborhood U in $\mathbb{R}^n$ such that $\Gamma \cap U = \{x\}$. Since U can be shrunk to a sphere with rational radius and whose center has rational coordinates, and since there is only a countable number of such spheres, we see that the subset Γ is necessarily countable. Because of this, and because $\mathbb{R}^n$ has a countable basis, any embedded or immersed zero-dimensional manifold of $\mathbb{R}^n$ must also be at most countable. As a matter of fact, we see that every zero-dimensional embedded manifold is discrete as well. This is not true for the immersed case since, for example, a countable dense subset of $\mathbb{R}^n$ satisfies the definition of an immersed manifold of dimension zero; such subsets are known to exist (e.g., the set of points with rational coordinates).

Conversely, since every point is a zero-dimensional elementary subset of $\mathbb{R}^n$ (it has the implicit equations $x^i - a^i = 0$, $i = 1, \ldots, n$), it follows that every discrete subset is an embedded zero-dimensional manifold.

(b) $m = n$. In this case it follows straightforwardly from the definitions that a subset Γ of $\mathbb{R}^n$ is elementary n-dimensional iff it is an open subset of $\mathbb{R}^n$. Hence an n-dimensional embedded or immersed manifold must be a union of open sets (i.e., it must be itself an open set).

(c) $m = 1$. In this case the manifolds are called curves or lines. For example, a straight line is an elementary one-dimensional subset, hence a curve. The subset of $\mathbb{R}^2$ whose points are characterized by

$$(x^1)^2 + (x^2)^2 = 1 \tag{1.19}$$

(i.e., the unit circle) is a curve as well. The subset $A = B \cup C$ of $\mathbb{R}^2$, where

$$B = \left\{(x^1, x^2) \mid x^1 \neq 0,\ x^2 - \sin\frac{1}{x^1} = 0\right\}$$
$$C = \{(x^1, x^2) \mid x^1 = 0\} \tag{1.20}$$

is an immersed curve in $\mathbb{R}^2$. In fact, if $b \in B$ and $b = (b^1, b^2)$ with $b^1 > 0$, the open half plane $x^1 > 0$ is a neighborhood of b whose intersection with A is characterized by $x^2 - \sin(1/x^1) = 0$, $x^1 > 0$, and it is elementary one-dimensional. A similar situation occurs for $b^1 < 0$. But if $c \in C$, the whole of $\mathbb{R}^2$ is an open neighborhood of c whose intersection with A consists of three elementary one-dimensional subsets, and if $c = (0, c^2)$ with $|c^2| \leq 1$, there is no neighborhood U of c such that $U \cap A$ is one elementary subset. (See Fig. 1.1, where we identify $\mathbb{R}^2$ with the intuitive E^2 endowed with a euclidean frame.)

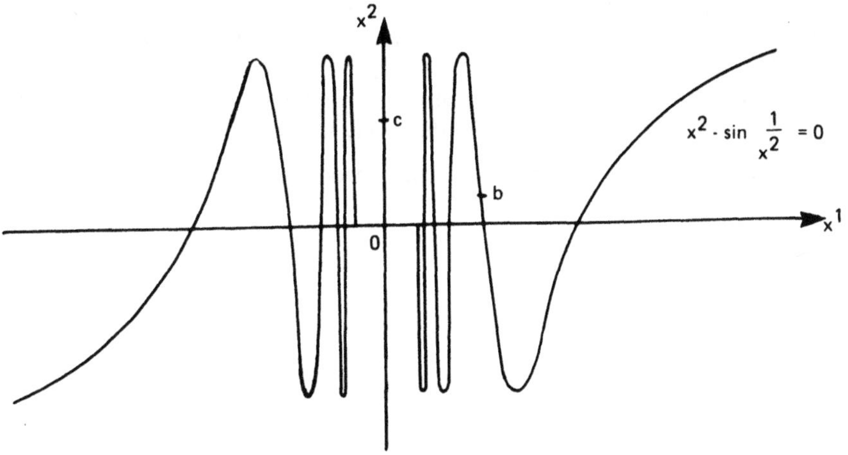

Fig. 1.1

(d) $m = 2$. In this case the manifolds are called surfaces. For instance, every plane is a surface. The spheres of $\mathbb{R}^3$ are surfaces. Other examples of surfaces are considered in Chap. 3.

(e) $m = n - 1$. In this case the manifolds are called hypersurfaces. For instance, the hyperplanes and the hyperspheres of $\mathbb{R}^n$ are hypersurfaces.

(f) Let $f: U \to \mathbb{R}^n$ be a diffeomorphism defined on the open subset $U \subseteq \mathbb{R}^m$, $m \leq n$ (see Definition 1.5). Then we shall prove

PROPOSITION 1.2 $f(U)$ is an embedded m-dimensional manifold in $\mathbb{R}^n$.

Proof: Let us consider an arbitrary point $x_0 \in f(U)$. Then there is a unique point $u_0 \in U$ such that $x_0 = f(u_0)$.

Denote by x^i ($i = 1, \ldots, n$) the coordinates in $\mathbb{R}^n$ and by u^α ($\alpha = 1, \ldots, m$) the coordinates in $\mathbb{R}^m$. Then we can represent f by

$$x^i = x^i(u^\alpha) \tag{1.21}$$

and since the points of U are nonsingular for f, there is a set $(i_1, \ldots, i_m)$ of distinct indices among $1, \ldots, n$ such that

$$\operatorname{rank} \left(\frac{\partial x^{i_\alpha}}{\partial u^\beta} \right)_{u=u_0} = m \qquad \alpha, \beta = 1, \ldots, m \tag{1.22}$$

Let us also consider the complementary set $\{j_1, \ldots, j_{n-m}\}$ of $\{i_1, \ldots, i_m\}$ in $\{1, \ldots, n\}$, and let us take the decomposition $\mathbb{R}^n = \mathbb{R}^m \times \mathbb{R}^{n-m}$

given by $(x^i) \longmapsto ((x^{i_\alpha}), (x^{j_\sigma}))$ $(\alpha = 1, \ldots, m; \sigma = 1, \ldots, n - m)$. The projection of this decomposition onto the first factor will be denoted by $p : \mathbb{R}^n \to \mathbb{R}^m$.

Then, $g = p \circ f$ is represented by $x^{i_\alpha} = x^{i_\alpha}(u^\beta)$, and in view of (1.22) we can apply the inverse function theorem, which yields an open neighborhood B of $(x_0^{i_\alpha})$ in $\mathbb{R}^m$, an open neighborhood A of u_0 in U, and a unique inverse $g^{-1} : B \to A$ defined by

$$u^\alpha = u^\alpha(x^{i_\beta}) \tag{1.23}$$

such that $u_0^\alpha = u^\alpha(x_0^{i_\beta})$, and

$$x^{i_\alpha} - x^{i_\alpha}(u^\alpha(x^{i_\beta})) = 0 \tag{1.24}$$

From the above we have that $f(A) = f \circ g^{-1}(B)$, and since $p \circ f = g$, this is just $f(U) \cap D$, where $D = B \times \mathbb{R}^{n-m}$ (i.e., an open subset of $\mathbb{R}^n$). Moreover, f(A) is the subset of D defined by the equations

$$x^{j_\sigma} - x^{j_\alpha}(u^\alpha(x^{i_\beta})) = 0 \qquad \sigma = 1, \ldots, n - m \tag{1.25}$$

and therefore f(A) is an elementary m-dimensional subset of $\mathbb{R}^n$. Hence f(U) is an embedded m-dimensional manifold in $\mathbb{R}^n$. Q.E.D.

It is important to note that the proof above can be applied to any $(u_0, f(u_0))$, $u_0 \in U$, for any differentiable function $f : U \to \mathbb{R}^n$ such that u_0 is not a singular point of f. This assures the existence of the neighborhoods A, B above, and proves that $f \mid A$ is a diffeomorphism. That is, we have proven

PROPOSITION 1.3 A differentiable function $f : U \to \mathbb{R}^n$, where U is an open subset of $\mathbb{R}^m$ $(m \leq n)$, is a local diffeomorphism iff f has no singular points in U.

Moreover, for such a function, $f : U \to \mathbb{R}^n$, the proof of Proposition 1.3 shows that we can write f(U) as a union of open neighborhoods which are m-dimensional elementary subsets. Since $\mathbb{R}^n$ has a countable basis, this union can also be taken as countable. This proves

PROPOSITION 1.4 Let $f : U \to \mathbb{R}^n$ be a differentiable function without singular points defined on the open subset $U \subset \mathbb{R}^m$ $(m \leq n)$. Then f(U) is an immersed m-dimensional manifold of $\mathbb{R}^n$.

These examples suggest

DEFINITION 1.9 Every diffeomorphism $f: U \to \mathbb{R}^n$ as in Proposition 1.2 is called an embedding of U in $\mathbb{R}^n$ and every differentiable map $f: U \to \mathbb{R}^n$ without singular points is called an immersion of U in $\mathbb{R}^n$.

(g) For the time being, the following will be the last example (the Descartes folium): Let

$$\Delta = \{(x^1, x^2) \in \mathbb{R}^2 \mid (x^1)^3 + (x^2)^3 - 3ax^1x^2 = 0\} \tag{1.26}$$

where a is a constant real number.

Δ is not an elementary curve because both derivatives of the left side of its equation vanish at the origin and the latter is a point of Δ. However, we can show that Δ is a curve immersed in $\mathbb{R}^2$. Indeed, using the substitution $x^2 = tx^1$, we get from (1.26) that $\Delta = f(\mathbb{R} - \{-1\})$, where f is the immersion of $\mathbb{R} - \{-1\}$ in $\mathbb{R}^2$ defined by

$$x^1 = \frac{3at}{1 + t^3} \qquad x^2 = \frac{3at^2}{1 + t^3} \tag{1.27}$$

The result announced then follows by Proposition 1.4.

Example (g) shows that systems of implicit equations with singular points may define immersed manifolds. But this is not always true.

Generally, if P is a subset of $\mathbb{R}^n$ which, except for some isolated points, satisfies the definition of a manifold, we shall say that P is a manifold with singular points, the singular points being those points where the manifold definition does not hold. We shall consider such manifolds only rarely, and then always with the necessary specification.

The equation of Δ given by (1.26) is algebraic (i.e., the vanishing of a polynomial). Generally, the manifolds (possibly with singular points) which are defined by algebraic equations are called algebraic manifolds and they are studied in an important special discipline: algebraic geometry.

Finally, let us also mention the following terminology: A manifold that is topologically compact is called a closed manifold; a manifold with no compact connected component is called an open manifold.

EXERCISES

1.14 Prove that the equation

$$(x^1)^2 - (x^2)^2 - [(x^1)^2 + (x^2)^2]^2 = 0$$

defines an immersed curve of $\mathbb{R}^2$ which is not embedded in $\mathbb{R}^2$. [Hint: Use the method of Example (g).]

1.15 Prove that the image of the mapping $f: \mathbb{R} \to \mathbb{R}^3$ defined by the equations

$$x^1 = t \quad x^2 = t^2 \quad x^3 = t^3$$

is an embedded curve of $\mathbb{R}^3$.

1.16 Prove that the quadratic cones are the only quadrics that are neither embedded nor immersed surfaces in $\mathbb{R}^3$. [Hint: The vertex does not belong to any elementary two-dimensional set.]

1.17 Consider the subset of $\mathbb{R}^2$ defined by the equation

$$(x^1)^2 - (x^2)^3 = 0$$

Prove that this set is neither an embedded nor an immersed curve. [Hint: It has a cusp at the origin.]

1.18 Let $M^m \subset \mathbb{R}^h$ and $N^n \subset \mathbb{R}^k$ be two embedded (immersed) manifolds of dimensions m and n, respectively. Prove that the cartesian product $M \times N$ is an embedded (immersed) $(m + n)$-dimensional submanifold of $\mathbb{R}^{h+k}$.

1.19 Prove that the sphere S^{n-1} is not the image of a diffeomorphism $\varphi: U \to \mathbb{R}^n$, where U is an open subset of an arithmetic space. [Hint: Use the fact that the sphere is compact.]

1.4 PARAMETERIZATIONS AND MAPS

In this section we provide another insight into the structure of a manifold, which is based on the following theorem:

THEOREM 1.3 Let M be an arbitrary embedded manifold in $\mathbb{R}^n$ and let x_0 be an arbitrary point of M. Then there is an open neighborhood V of x_0 in M such that V is the image of some diffeomorphism.

Proof: Suppose that M is m-dimensional. Then x_0 has an open neighborhood U in $\mathbb{R}^n$ such that $U \cap M = W$, which is an open neighborhood of x_0 in M, is elementary m-dimensional, and has the implicit equations

$$F_\sigma(x^1, \ldots, x^n) = 0 \quad \sigma = 1, \ldots, n - m \tag{1.28}$$

where F_σ are differentiable on U and

$$\text{rank}\left(\frac{\partial F_\sigma}{\partial x^i}\right) = n - m \tag{1.29}$$

Considering the map $F: U \to \mathbb{R}^{n-m}$ defined by the function F_σ, we can apply the implicit function Theorem 1.1 at the point x_0. This provides a decomposition $\mathbb{R}^n = \mathbb{R}^{n-m} \times \mathbb{R}^m$, $x_0 = (x_0', x_0'')$ $(x_0' \in \mathbb{R}^{n-m}, x_0'' \in \mathbb{R}^m)$, two open neighborhoods A of x_0' in $\mathbb{R}^{n-m}$ and B of x_0'' in $\mathbb{R}^m$, and a unique function $\varphi : B \to A$, such that $A \times B \subseteq U$, $x_0' = \varphi(x_0'')$, $F(\varphi(y), y) = 0$ for every $y \in B$.

Now, let us denote by $\Phi : B \to \mathbb{R}^n$ the differentiable map that sends every $y \in B$ to $(\varphi(y), y) \in \mathbb{R}^n$. This map has no singular points. Then we have $\Phi(B) = \varphi(B) \times B \subseteq W$, and because of the uniqueness part of the implicit function theorem, we get $\Phi(B) = W \cap (R^{n-m} \times B)$. It follows that $\Phi(B)$ is an open neighborhood of x_0 in W (i.e., in M as well), and we see that Φ is the diffeomorphism given by Theorem 1.3. Q.E.D.

Theorem 1.3 suggests

DEFINITION 1.10 If M is an embedded manifold of $\mathbb{R}^n$, any embedding $\Phi : U \to \mathbb{R}^n$, where U is an open subset of some arithmetic space whose image $\Phi(U)$ is an open subset $V \subseteq M$, is called a parameterization of M. In this case the inverse mapping $\Phi^{-1} = \Psi : V \to U$ is called a (local) map or chart on M.

Let us introduce some more terminology related to these notions. With the notation above, if $x_0 \in V$, Φ is called a parameterization of M at x_0 and Ψ is a map or chart at x_0, U is called either the domain of the parameterization Φ or the range of the map Ψ, and V is the domain of the map Ψ and the range of the parameterization Φ. The arithmetic coordinates u^α in U are called either local parameters or local coordinates on M (at x_0). The family of all the charts of M forms the atlas of M.

Let us consider two parameterizations Φ_a (a = 1, 2) of M. We shall use for them all the symbols mentioned above, endowed with the index a.

If $V_1 \cap V_2 = W \neq \emptyset$, W is an open subset of M, and we get a homeomorphism

$$h_{21} : \Psi_1(W) \to \Psi_2(W) \tag{1.30}$$

defined by

$$h_{21} = \Psi_2 \circ \Phi_1 = \Phi_2^{-1} \circ \Phi_1 \tag{1.31}$$

This will be called the transition function or the change of parameters between the two parameterizations.

Because h_{21} is a homeomorphism, the domains U_a of the two parameterizations must be open sets in arithmetic spaces of the same dimension. Hence, since at every point $x_0 \in M$ there is a parameterization with domain in $\mathbb{R}^m$, where m is the dimension of M (this was shown in the proof of Theorem 1.3), we see that the number of the local parameters is always m.

The coordinate expression of a parameterization is

$$x^i = x^i(u^\alpha) \qquad i = 1, \ldots, n; \quad \alpha = 1, \ldots, m \tag{1.32}$$

(here each x^i depends on all the u^α) or, vectorially,

$$\underline{x} = \underline{x}(u^\alpha) \tag{1.33}$$

and these are called the local parametric equations of the manifold M.

We note that Theorem 1.3 yields particular parameterizations, which are of the form

$$x^{i_\alpha} = x^{i_\alpha} \qquad x^{j_\sigma} = x^{j_\sigma}(x^{i_\alpha}) \tag{1.34}$$

where $\alpha = 1, \ldots, m$; $\sigma = 1, \ldots, n - m$, and $\{i_\alpha\}$, $\{j_\sigma\}$ are two complementary subsets of $\{1, \ldots, n\}$. For such parameterizations, (1.34) are called the local explicit equations of M. We are also reminded that any parametric equations must express a diffeomorphism and, particularly, must have no singular points.

The coordinate expression of a change of parameters is of the form

$$u_2^\alpha = u_2^\alpha(u_1^\beta) \qquad \alpha, \beta = 1, \ldots, m \tag{1.35}$$

Recalling that (1.35) represents (1.31), and by taking into account that Φ_a are diffeomorphisms, and that Φ_2^{-1} is given locally by the inverse function theorem, we see that (1.35) are differentiable functions and that a change of parameters is actually a diffeomorphism of open subsets of $\mathbb{R}^m$. Note also that, by the discussion in Sec. 1.2, this implies that the jacobian of (1.35) does not vanish, and that this nonvanishing condition is also sufficient for (1.35) to define a diffeomorphism locally.

As a matter of fact, Theorem 1.3 shows how to go over, locally, from the implicit equations of an embedded manifold to parametric equations via explicit equations. On the other hand, in Example (f) of Sec. 1.3, we saw how to go over locally from parametric equations to implicit equations, via explicit equations. Hence the three types of equations mentioned are locally equivalent, and this will be used in the following considerations.

Namely, by Theorem 1.3, we see that an embedded manifold admits a (countable) open covering $\{V_\lambda\}$ such that each V_λ is the domain of a map $\Psi_\lambda : V_\lambda \to U_\lambda \subseteq \mathbb{R}^m$, and that any two maps are related by means of some transition function which is a diffeomorphism.

Conversely, if $M \subseteq \mathbb{R}^n$ and if M can be covered by an atlas $\{V_\lambda\}$ as above, then, by "going over from parametric to implicit equations," we can see that M is an embedded manifold in $\mathbb{R}^n$. Hence we obtain a new viewpoint of embedded differentiable manifolds.

Parameterizations and maps can also be used for immersed manifolds. Indeed, if $x_0 \in N$ and N is an immersed manifold, x_0 belongs to some elementary sets Γ_α and for every one of these Γ_α we may use parametric equations. However, now there is no transition function between parameterizations of different sets Γ_α at the same point x_0 if those Γ_α have no open intersection.

This new understanding of differentiable manifolds leads to important developments which we shall mention here only in passing. First, it indicates the way to abstract differentiable manifolds, which are defined as topological Hausdorff spaces admitting a (countable) atlas of the type mentioned above. We must note, however, that by an important theorem of Whitney, every such abstract manifold can be embedded in some space $\mathbb{R}^n$ for sufficiently large n.

Second, if we ask simply that the maps Ψ_λ be homeomorphisms, and that the transition functions of the atlas be homeomorphisms (i.e., we impose no differentiability conditions), we obtain the topological manifolds or manifolds of class C^0. The study of C^0 manifolds is a very important aspect of topology, but we will not discuss them in this book.

Let us make a few additional comments about differentiable manifolds.

As we have seen, the manifolds are characterized by local equations. Hence we must distinguish between local and global properties of the manifold. The local properties are those which are related to a neighborhood of a point on one elementary m-dimensional subset (i.e., properties related to the range of a parameterization). The global properties are those which are related to the whole manifold.

Correspondingly, we distinguish between local and global differential geometry. A general method that yields global properties can be described as follows: First, one looks for properties valid in domains of charts, and next, one tries to glue up the results related to domains of charts by using the transition functions.

There is one more problem related to the coordinate representation of a manifold. Namely, if we are interested in properties of the manifold and not of its representation, we must verify that the properties obtained are invariant under changes of the analytic representation of the manifolds. In particular, if we use parameterizations, which is usually the case, we must verify invariance under changes of parameters. The invariant properties are also called geometric properties of the manifold, while the invariant quantities are called invariants of the manifold.

Another problem is that of going over from $\mathbb{R}^n$ to an arbitrary euclidean space E^n. It is easy to see that the conditions which enter into the definitions of embedded and immersed manifolds in $\mathbb{R}^n$ are invariant by coordinate transformations of the form (1.11). Hence we can define embedded and immersed manifolds in E^n as subsets $M \subseteq E^n$ which can be identified with manifolds in $\mathbb{R}^n$ by introducing a euclidean frame in E^n.

In later chapters, we will be interested in manifolds of E^n. One additional item should be emphasized: that the geometric properties and the

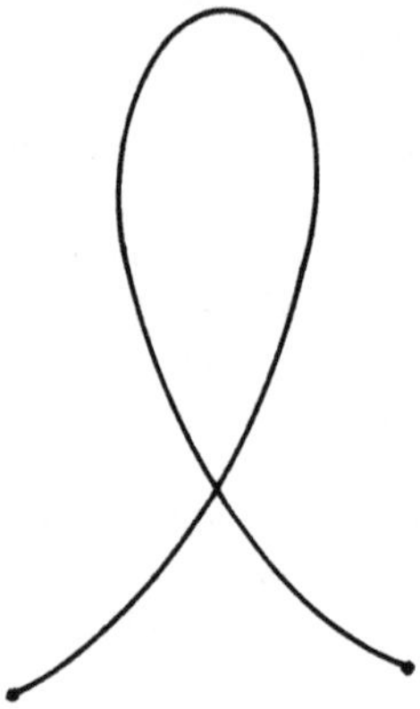

Fig. 1.2

invariants of a manifold are, in this case, those which are invariant under both change of parameters (or of the coordinate representation of the manifold) and coordinate transformation (1.11).

Finally, let us note that our definitions of curves, surfaces, and manifolds do, however, leave out some objects that our intuition would tell us to include. This is the case, for example, for the "curve" of Fig. 1.2, where the two end points do not satisfy the curve definition. This difficulty can be overcome by defining manifolds with boundaries. These can be introduced by Definitions 1.6, 1.7, and 1.8, where we now add, as new m-dimensional elementary subsets, subsets defined by

$$\Gamma = \{x \in U \mid F_\sigma(x) = 0, \ F_{n-m+1}(x) \geq 0\}$$

The notation is as in Definition 1.6, and the differentiable functions F_σ, F_{n-m+1} have to define a mapping without singular points. However, in this book we will not study such manifolds.

EXERCISES

1.20 Consider the map ψ of Exercise 1.8 and the similar map $\psi' : \mathbb{R}^2 \to \mathbb{R}^3$ which is obtained by replacing the point (0, 0, 1) by (0, 0, -1) in the definition of ψ. Show that these two maps are parameterizations of the sphere S^2 whose ranges cover S^2. Write down the transition functions (change of parameters) between these two parameterizations.

1.21 Find a covering of the sphere S^2 by ranges of local explicit equations. Write the corresponding transition functions.

1.22 Write the linear map defined by the differential of the transition functions between two parameterizations at an arbitrary point.

1.23 Let M be an m-dimensional differentiable manifold embedded in $\mathbb{R}^n$, and let U be the domain of a parameterization Φ of M at $x_0 = \Phi(u_0) \in M$. Define a transformation $h: V \to \mathbb{R}^m$, where V is an open subset of U containing u_0, by equations of the form

$$u_2^\alpha = u_2^\alpha(u_1^\beta) \qquad \alpha, \beta = \ , \ldots, m \tag{*}$$

Prove that an open neighborhood W of x_0 on M exists such that the u_2^α of (*) are local parameters on W, if and only if the jacobian

$$\det \left[\frac{\partial u_\alpha^2}{\partial u_\alpha^1} \right]_0 \neq 0$$

at u_0. [Hint: Use the inverse function theorem.]

1.24 The proportionality relation of vectors in $\mathbb{R}^{n+1}$ is an equivalence relation ρ. Take the following set of equivalence classes: $P^n = (\mathbb{R}^{n+1} \setminus \{0\})/\rho$; this is called real n-dimensional projective space. Prove that P^n is an abstract n-dimensional differentiable manifold. [Hint: Let $U_i \subset P^n$ be the subset with $x^i \neq 0$, where x^i ($i = 1, \ldots, n+1$) are the coordinates in $\mathbb{R}^{n+1}$. Define $h_i: U_i \to \mathbb{R}^n$ by $h_i([x^j]_\rho) = (y^j)$, where $y^j = x^j/x^i$ ($j = 1, \ldots, n+1$; $j \neq i$). Prove that these n + 1 functions define an atlas with differentiable transition functions.]

1.25 Prove in detail that if a subset M of a euclidean space E^n is an embedded (immersed) differentiable manifold with respect to one frame of E^n, this will hold for every other frame as well.

1.26 Prove that a usual (two-dimensional) closed hemisphere is an embedded two-dimensional differentiable manifold with boundary in $\mathbb{R}^3$.

1.5 THE TANGENT SPACE OF A MANIFOLD

This section deals with the problem of giving a rigorous and general definition of a concept corresponding to the intuitive tangent line to a curve and tangent plane to a surface. The intuitive notions provide a local approximation of a curve by a straight line and of a surface by a plane, which shows the importance of the problem under consideration. We also mention that, as is well known from elementary calculus, the tangent line of a curve is intimately related to differentiability.

Let us begin by introducing the important notion of a path. Generally, a path in a topological space T is either a continuous function $f:(a,b) \to T$, where (a,b) is an open interval in $\mathbb{R}$, or a continuous function $f:[a,b] \to T$, where $[a,b]$ is a closed interval in $\mathbb{R}$. In the second case $f(a)$ is the origin and $f(b)$ is the endpoint of the path, and if $f(a) = f(b)$ the path is said to be closed. Equivalently, a closed path can be thought of as a path defined by a continuous periodic function $f:\mathbb{R} \to T$ with a period $b - a$. We are emphasizing the fact that the path is the function f and not its range only.

Now consider the particular case $T = \mathbb{R}^n$ with its euclidean structure. A path in $\mathbb{R}^n$ is said to be differentiable if the corresponding function f is differentiable, in which case it is said to be regular if the function f has no singular points. From the results of Example (f) of Sec. 1.3 it follows that the range of a regular differentiable path is an immersed curve in $\mathbb{R}^n$, which will be called the curve defined by the paths f, and which must not be confused with f itself.

Let $f:(a,b) \to \mathbb{R}^n$ be a path in $\mathbb{R}^n$, $t_0 \in (a,b)$, and $x_0 = f(t_0)$. Then elementary physics suggests defining the velocity vector of f at t_0 by

$$\underline{v}(t_0) = \lim_{t \to t_0} \frac{f(t) - f(t_0)}{t - t_0} \tag{1.36}$$

whenever this limit exists.

DEFINITION 1.11 In differential geometry, the velocity vector $\underline{v}(t_0)$ defined by (1.36) is called the tangent vector of the path f at t_0.

PROPOSITION 1.5 If f is a differentiable path, the tangent vector exists and it is given by $\underline{v}(t_0) = df/dt|_0$.

Proof: By applying to f the Taylor formula (1.18) we get

$$f(t) - f(t_0) = \left.\frac{df}{dt}\right|_0 \cdot (t - t_0) + o(1)$$

where o(1) are terms of "second degree" with respect to $t - t_0$. Dividing by $t - t_0$, and taking $t \to t_0$, we obtain the stated result.

Let us note that the velocity vector is 0 at the singular points of the path only. This vector depends on the path and not only on the curve defined by that path, and it has a local character; that is, it depends only on the restriction of f to an arbitrary open neighborhood of t_0. But, for example, if $f(t_0) = f(t_1)$ for $t_0 \neq t_1$, we generally have $\underline{v}(t_0) \neq \underline{v}(t_1)$. It is also worth noting that one has

$$\underline{v}(t_0) = f'_{t_0}(1) \tag{1.37}$$

where $f'_{t_0}:\mathbb{R} \to \mathbb{R}^n$ is the derivative of f as introduced by Definition 1.3. This follows directly from formula (1.14).

We now consider manifolds.

DEFINITION 1.12 Let $M \subseteq \mathbb{R}^n$ be an embedded differentiable manifold, and $f: (a,b) \to \mathbb{R}^n$ a path. Then f is said to be a <u>path in</u> M if range $f \subseteq M$.

Note that the mapping $f: (a,b) \to M$ is continuous. Indeed, since M is embedded, it has the induced topology, and if $D \subseteq M$ is open we have $D = D' \cap M$, where D' is open in $\mathbb{R}^n$. Then $f^{-1}(D) = f^{-1}(D')$, which is open because f is continuous.

Now, let Γ be an elementary m-dimensional subset of the open set $U \subseteq \mathbb{R}^n$ and $x_0 \in \Gamma$. Suppose that the implicit equations of Γ are

$$F_\sigma(x^1, \ldots, x^n) = F_\sigma(x^i) = 0 \qquad \sigma = 1, \ldots, n-m \tag{1.38}$$

Then a path is in Γ iff it satisfies the equations

$$F_\sigma(x^i(t)) \equiv 0 \tag{1.39}$$

where $x^i = x^i(t)$ are the equations of the path, and we say that the path passes through x_0 if $x_0^i = x^i(t_0)$ for some value of the <u>parameter</u> t.

By differentiating (1.39) we see that the velocity vector $\underline{v}$ of such a path satisfies the equations

$$\sum_{i=1}^{n} \left.\frac{\partial F_\sigma}{\partial x^i}\right|_0 \cdot v^i = 0 \tag{1.40}$$

where $v^i = dx^i/dt\,|_0$ are the coordinates of the vector $\underline{v}$.

In view of Definition 1.6, rank $(\partial F_\sigma/\partial x^i)_0 = n - m$, whence (1.40) defines an m-dimensional linear subspace of $\mathbb{R}^n$, which we shall denote by $T_{x_0}(\Gamma)$.

To proceed with our discussion of $T_{x_0}(\Gamma)$, let us consider a parameterization $\Phi: U \to \mathbb{R}^n$ of Γ at x_0 given by

$$x^i = x^i(u^\alpha) \qquad i = 1, \ldots, n; \ \alpha = 1, \ldots, m \tag{1.41}$$

Then we clearly have

$$F_\sigma(x^i(u^\alpha)) \equiv 0 \tag{1.42}$$

and by differentiating this relation

$$\sum_{i=1}^{n} \left.\frac{\partial F_\sigma}{\partial x^i}\right|_0 \cdot \left.\frac{\partial x^i}{\partial u^\alpha}\right|_0 = 0 \tag{1.43}$$

where u_0^α are the local coordinates of x_0 on Γ.

Hence the vectors

$$\underline{x}^0_\alpha = \left.\frac{\partial x}{\partial u^\alpha}\right|_0 \qquad \alpha = 1, \ldots, m \tag{1.44}$$

which are m linearly independent vectors, define a basis of the space $T_{x_0}(\Gamma)$.

The coordinates in $T_{x_0}(\Gamma)$ of the vector $\underline{v}$ of (1.40) with respect to the basis (1.44) can be calculated as follows. It is clear that for some $(c,d) \subset (a,b)$ with $t_0 \in (c,d)$, range $f|(c,d) \subseteq$ range Φ. Hence we get a mapping $g:(c,d) \to U$ defined by $g = \Phi^{-1} \circ f$, which will be represented by

$$u^\alpha = u^\alpha(t) \qquad \alpha = 1, \ldots, m \tag{1.45}$$

Moreover, the mapping g is clearly differentiable, which follows by using the transition relations between Φ and a parameterization Ψ of Γ at x_0 by explicit equations. Indeed, we have then $g = (\Phi^{-1} \circ \Psi) \circ (\Psi^{-1} \circ f)$, where the two parentheses are differentiable.

Hence $f = \Phi \circ g$, which means that along (c,d) the path f has the equations

$$x^i = x^i(u^\alpha(t)) \tag{1.46}$$

which is the composition of (1.41) and (1.45), and its velocity vector at t_0 is

$$\underline{v} = \frac{dx}{dt} = \sum_{\alpha=1}^{m} \left.\frac{du^\alpha}{dt}\right|_0 \cdot \underline{x}^0_\alpha \tag{1.47}$$

which yields the coordinates required.

It is now easy to see that every vector $\underline{w} \in T_{x_0}\Gamma$ is the velocity vector of some path on Γ through x_0. Namely, if $\underline{w} = \Sigma_{\alpha=1}^{m} w^\alpha \underline{x}^0$, then $u^\alpha = w^\alpha t$ define, by (1.46), the path required: $x^i = x^i(w^\alpha t)$.

Hence $T_{x_0}\Gamma$ is exactly the set of all the velocity vectors of the paths of Γ through x_0. Therefore, $T_{x_0}\Gamma$ has an invariant meaning, and does not depend on the analytic representation of Γ.

The following facts are interesting as well:

PROPOSITION 1.6 In the notation above, we have $T_{x_0}\Gamma = \Phi'_{u_0}(\mathbb{R}^m)$, where Φ is a parameterization at x_0, $x_0 = \Phi(u_0)$, and $\Phi'_{u_0}: \mathbb{R}^m \to \mathbb{R}^n$ is the derivative mapping of Φ.

Proof: Let Φ be defined by (1.41). Then, by formula (1.15), Φ'_{u_0} has the equations

$$\eta^i = \frac{\partial x^i}{\partial u^\alpha} \xi^\alpha$$

By applying these equations to the vectors of the natural basis of $\mathbb{R}^m$, whose coordinates are $\xi^\alpha = \delta^\alpha_\beta$ ($\delta^\alpha_\beta = 1$ if $\alpha = \beta$; $\delta^\alpha_\beta = 0$ if $\alpha \neq \beta$; $\alpha, \beta = 1, \ldots, m$), we see that the image space $\Phi'_{u_0}(\mathbb{R}^m)$ has the same basis $\underline{x}^0_\alpha$ as $T_{x_0}\Gamma$. Q.E.D.

PROPOSITION 1.7 If U is an open subset of $\mathbb{R}^n$ and $x_0 \in U$, one has $T_{x_0}U = \mathbb{R}^n$.

Proof: Let us take for U the parameterization Φ defined by the inclusion $U \subseteq \mathbb{R}^n$. This parameterization has the equations $x^i = x^i$ $(i = 1, \ldots, n)$, and the corresponding basis (1.44) is the natural basis of $\mathbb{R}^n$. Q.E.D.

Clearly, the whole preceding discussion holds for any embedded manifold M, and working with neighborhoods of x_0 in M which are elementary sets Γ, we get a well-defined m-dimensional linear space $T_{x_0}M$.

Now we introduce some corresponding terminology.

DEFINITION 1.13 Let M be a differentiable manifold embedded in $\mathbb{R}^n$ and $x_0 \in M$. Then every velocity vector of a path of M through x_0 is called a tangent vector of M at x_0. The linear space $T_{x_0}M$ is called the tangent space of M at x_0. Its basis (1.44) is called the natural basis of the parameterization Φ (or of the map Φ^{-1}), and the coordinates of a tangent vector with respect to such a basis are called natural or internal coordinates of the vector.

From the viewpoint of the ambient space $\mathbb{R}^n$, it is sometimes more interesting to consider

DEFINITION 1.14 The m-dimensional plane $\pi_{x_0}M$ of euclidean space $\mathbb{R}^n$ which passes through the point x_0 and whose vectors are the vectors of $T_{x_0}M$ is called the tangent plane of M at x_0.

The tangent plane $\pi_{x_0}M$ should not be confused with the tangent space $T_{x_0}M$.

If $m = 1$ (M is a curve), we speak of the tangent line. If $m = n - 1$, we have the tangent hyperplane.

The computations already performed reveal to us how to represent $\pi_{x_0}M$ analytically. Namely, if we take the implicit equations (1.38) in a neighborhood of x_0, then, in view of (1.40), $\pi_{x_0}M$ is defined by the following system of linear equations:

$$\sum_{i=1}^{n} \left.\frac{\partial F_\sigma}{\partial x^i}\right|_0 \cdot (x^i - x_0^i) = 0 \qquad (1.48)$$

If we take a parameterization Φ at x_0, then, in view of the existence of the natural basis (1.44), $\pi_{x_0}M$ is defined by the parametric vector equations

$$\underline{x} = \underline{x}_0 + \sum_{\alpha=1}^{m} \lambda^{\alpha} \underline{x}_{\alpha}^{0} \tag{1.49}$$

where the λ^{α} are parameters. For explicit equations of M, we can particularize either (1.48) or (1.49).

In the case of a curve (m = 1), a parameterization is simply a nonsingular differentiable path through x_0, and the tangent line is the line through x_0 whose direction is given by the velocity vector of the path. But before taking the limit, the vector $f(t) - f(t_0)$ of (1.36) can be seen as the direction of a secant of the curve, that is, the line joining x_0 to another point x of the curve. Hence one can say (and this terminology can, in fact, be made rigorous, which is not done here) that the tangent line is the limit of the secant $x_0 x$ when $x \to x_0$ on the curve. This was the origin of the names tangent space and plane. (See Fig. 1.3 as an illustration.)

Furthermore, since our manifolds are in $\mathbb{R}^n$, it is natural to consider

DEFINITION 1.15 Any vector of $\mathbb{R}^n$ that is orthogonal to the tangent space $T_{x_0}M$ (i.e., to all its vectors) is called a normal vector of M at x_0. The linear (n - m)-dimensional vector space generated by these vectors is called the normal space of M at x_0, and will be denoted by $N_{x_0}M$. The (n - m)-dimensional plane of the euclidean space $\mathbb{R}^n$ which passes through

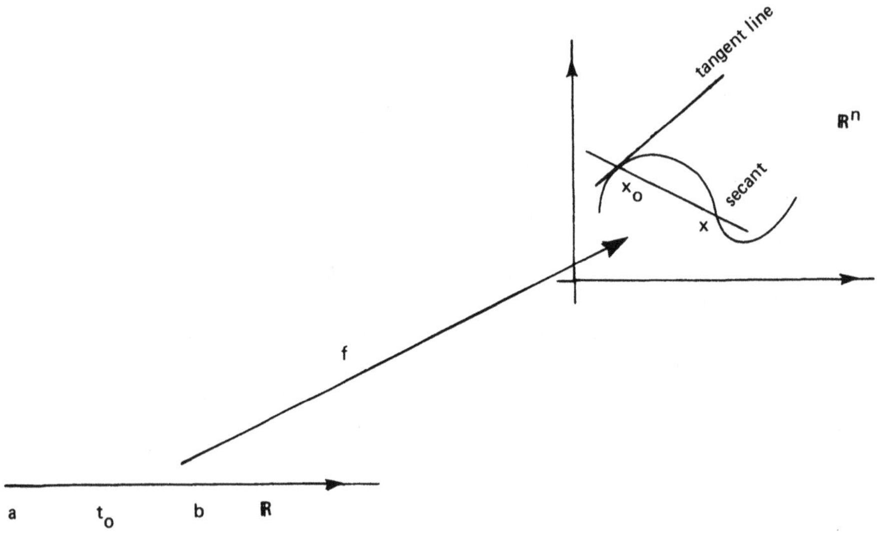

Fig. 1.3

x_0 and whose vectors are the normal vectors is called the normal plane of M at x_0, and it is denoted by $\nu_{x_0} M$. Any line through x_0 and contained in $\nu_{x_0} M$ is called a normal line of M at x_0.

For example, for a hypersurface, $\nu_{x_0} M$ is a straight line and we simply speak of the normal line of M at x_0.

The analytic representation of $\nu_{x_0} M$ follows easily from (1.48) and (1.49). That is, if we have implicit equations of M at x_0, (1.48) shows that $\nu_{x_0} M$ is generated by the vectors grad F_σ ($\sigma = 1, \ldots, n - m$), and it has the parametric equations

$$x = x_0 + \sum_{\sigma=1}^{n-m} \lambda^\sigma \operatorname{grad} F_\sigma \tag{1.50}$$

If we have a parameterization of M at x_0, we must first find a complementary orthogonal basis to the natural bases $\underline{x}^0_\alpha$.

We end this section with some already familiar remarks. First, tangent and normal spaces can be considered for immersed manifolds as well as embedded manifolds. But in this case, x_0 may belong to more than one elementary subset Γ, and for each one of them we have a corresponding tangent and normal space, which we define as above.

Second, from the definition of the velocity vectors we see that these are invariant by orthogonal transformations of the form (1.11). This fact implies the same invariance property for all tangent and normal spaces and planes. Therefore, the notions of a tangent and a normal actually have a euclidean character, and can be used for manifolds in an arbitrary euclidean space.

EXERCISES

1.27 Consider a differentiable manifold M and $x_0 \in M$. Let x(t) define a path of M through x_0. Then the vector $\underline{w} = (d^2\underline{x}/dt^2)_0$ is called the acceleration of the path at x_0. Prove that if the velocity of the path at x_0 vanishes $\underline{w}$ belongs to $T_{x_0} M$, but if the velocity is not zero, $\underline{w}$ may not belong to $T_{x_0} M$.

1.28 Let α be an h-plane of $\mathbb{R}^n$. Prove that, for any $x \in \alpha$, the tangent plane $\pi_x \alpha = \alpha$.

1.29 Consider the unit sphere S^n as an embedded manifold of $\mathbb{R}^{n+1}$. Prove that, for any $x \in S^n$, the tangent space $T_x S^n$ is the subspace consisting of the vectors of $\mathbb{R}^{n+1}$ that are orthogonal to $\underline{x}$. Deduce that the tangent plane and the normal line of S^n at x as defined here, coincide with the classical tangent plane and normal line of the sphere.

1.30 Write the natural tangent basis for an explicit parameterization of the unit sphere S^n at an arbitrary point.

1.31 Prove that the tangent plane and the normal line to a quadric surface as defined here coincide with the tangent plane and normal line of the quadric as defined in solid analytic geometry.

1.32 Let $P = M \times N$ be the cartesian product manifold of Exercise 1.18. Prove that for any point $p = (x, y)$ $(x \in M,\ y \in N)$ of P there is an invariant isomorphism

$$T_pP \approx T_xM \oplus T_yN$$

where $\oplus$ denotes the direct sum of linear spaces.

1.33 Show that the Descartes folium [Example (g), Sec. 1.3] has two tangent lines at the origin.

1.34 Let M be an embedded differentiable manifold of the euclidean space E^n, and $x \in M$. Show that the tangent plane π_xM does not depend on the choice of the frame in E^n.

1.6 DIFFERENTIABLE MAPPINGS OF MANIFOLDS

Differentiable paths are examples of mappings between differentiable manifolds, and in this section we consider the corresponding general case.

Let M be an m-dimensional differentiable manifold in $\mathbb{R}^n$ and M' an m'-dimensional manifold in $\mathbb{R}^{n'}$. A mapping $\varphi : M \to M'$ is defined as usual, that is, as a set-theoretic function between the two sets of points M and M'. We want to see how such a mapping can be studied in this particular situation.

Let us denote by $i' : M' \subseteq \mathbb{R}^{n'}$ the corresponding immersion or embedding. Then $\psi = i' \circ \varphi$ sends M into $\mathbb{R}^{n'}$ and the following definition makes sense:

DEFINITION 1.16 The mapping $\varphi : M \to M'$ between the two manifolds is said to be differentiable (or C^k, C^ω) if the associated mapping $\psi = i' \circ \varphi : M \to \mathbb{R}^{n'}$ is differentiable (or C^k, C^ω).

PROPOSITION 1.8

(a) Any immersion $i : M \to \mathbb{R}^n$ is differentiable in the sense of Definition 1.16.

(b) If $\varphi : M \to M'$, $\psi : M' \to M''$ $(M \subseteq \mathbb{R}^n,\ M' \subseteq \mathbb{R}^{n'},\ M'' \subseteq \mathbb{R}^{n''})$ are differentiable, then $\psi \circ \varphi$ is differentiable as well.

Proof:

(a) In this case the relevant composite function is i itself, which is an inclusion map and therefore differentiable.

(b) By the hypotheses, φ extends to a differentiable map $\Phi : U \to \mathbb{R}^{n'}$ and ψ extends similarly to $\Psi : U' \to \mathbb{R}^{n''}$, where U and U' are open neighborhoods of M and M', respectively.

Hence $\psi \circ \varphi$ extends to $\Psi \circ \Phi$, which proves the assertion of the proposition.

Now suppose that we consider only embedded submanifolds, and let $\varphi : M \to M'$ be a differentiable mapping as in Definition 1.16. Take $x_0 \in M$, $y_0 = \varphi(x_0) \in M'$, and let $\Phi : U \to \mathbb{R}^n$ and $\Phi' : U' \to \mathbb{R}^{n'}$ be parameterizations at x_0 and y_0, respectively. Then $U \subseteq \mathbb{R}^m$, $U' \subseteq \mathbb{R}^{m'}$, and we shall denote by x^i, x'^j, u^α, and u'^β the coordinates in $\mathbb{R}^n$, $\mathbb{R}^{n'}$, $\mathbb{R}^m$, and $\mathbb{R}^{m'}$, respectively. Figure 1.4 presents this situation.

Since $\Phi'(U')$ is open in M', we have $\Phi'(U') = M' \cap D'$, where D' is open in $\mathbb{R}^{n'}$. Next, since $\psi = i' \circ \varphi$ is differentiable, it is also continuous, and we have that

$$\varphi^{-1}(\Phi'(U')) = \psi^{-1}(D') \cap M$$

is open in M. Hence $V = \Phi(U) \cap \varphi^{-1}(\Phi'(U'))$ is open in $\Phi(U)$ and $\Phi^{-1}(V)$ is open in U.

We can therefore define an associated mapping

$$\tilde{\varphi}_{U,U'} = \tilde{\varphi} : \Phi^{-1}(V) \to U' \tag{1.51}$$

by $\tilde{\varphi} = \Phi'^{-1} \circ \varphi \circ \Phi$, and such functions $\tilde{\varphi}$ will be called <u>local analytic representations</u> of φ. Of course, such representations depend on the parameterization chosen.

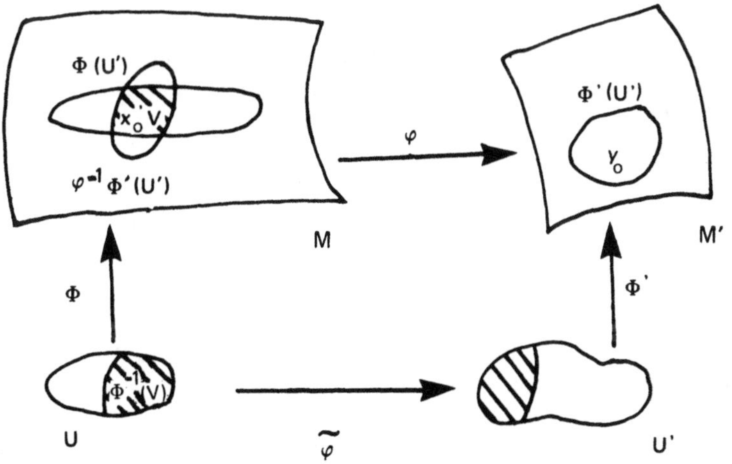

Figure 1.4

The function $\tilde{\varphi}$ can be written as

$$u'^{\beta} = u'^{\beta}(u^{\alpha}) \tag{1.52}$$

and we get

PROPOSITION 1.9 In the notation above, the local analytic representations $\tilde{\varphi}$ are differentiable, and they define the mapping φ uniquely.

Proof: The second assertion is obvious since we have (on corresponding domains) $\varphi = \Phi' \circ \tilde{\varphi} \circ \Phi^{-1}$. For the first assertion, it suffices to prove the differentiability of $\tilde{\varphi}$ at $\Phi^{-1}(x_0) = u_0$.

If Φ and Φ' are parameterizations given by explicit equations, where the parameters are $(x^{i_1}, \ldots, x^{i_m})$ and $(x'^{j_1}, \ldots, x'^{j_{m'}})$, respectively, then the equations (1.52) follow directly from the coordinate expression of $\psi = i' \circ \varphi$, and therefore define a differentiable mapping. The general case follows from this particular case because the transition functions between arbitrary local parameters and explicit parameters are differentiable functions. Q.E.D.

If we want to use the previous results for immersed manifolds, we have to consider them for every elementary subset Γ that contains some point $x_0 \in M$ and which is sent into some $\Gamma' \subseteq M'$.

Again let $\varphi: M \to M'$ be a differentiable mapping between two embedded submanifolds. We shall use all the anterior notation for it. Let $\underline{v} \in T_{x_0}M$ and suppose that $\underline{v}$ is the velocity vector of some path $f: (a, b) \to \mathbb{R}^n$ such that $x_0 = f(t_0)$ $[t_0 \in (a,b)]$. Then $f' = i' \circ \varphi \circ f: (a,b) \to \mathbb{R}^{n'}$ defines a path on M' through y_0; let $\underline{v}' \in T_{y_0}M'$ be the velocity vector of this path on M'. We shall prove

PROPOSITION 1.10 The vector $\underline{v}'$ above depends only on φ and $\underline{v}$, and not on the concrete choice of the path f. The correspondence $\varphi'_{x_0}: T_{x_0}M \to T_{y_0}M'$ defined by $\varphi'_{x_0}(\underline{v}) = \underline{v}'$ in a linear mapping.

Proof: Let us introduce parameterizations Φ at x_0 and Φ' at y_0. Then we have for f the local analytic representation (1.45), and we have for $\underline{v}$ the decomposition (1.47). For φ we shall have the representation (1.52), whence, for the path f', we have the local analytic representation

$$u'^{\beta} = u'^{\beta}(u^{\alpha}(t)) \tag{1.53}$$

Therefore, the expression of the vector $\underline{v}'$ with respect to the natural basis of the parameterization Φ' is

$$\underline{v}' = \sum_{\beta=1}^{m'} \left.\frac{du'^{\beta}}{dt}\right|_0 \cdot \underline{x}'^0_{\beta} = \sum_{\beta=1}^{m'} \sum_{\alpha=1}^{m} \left.\frac{\partial u'^{\beta}}{\partial u^{\alpha}}\right|_0 \cdot \left.\frac{du^{\alpha}}{dt}\right|_0 \underline{x}'^0_{\beta} \tag{1.54}$$

and this proves the assertions of Proposition 1.10.

Note that if we use the coordinates of the vectors with respect to natural bases of parameterizations, φ'_{x_0} is represented by the equations

$$v'^{\beta} = \sum_{\alpha=1}^{m} \left.\frac{\partial u'^{\beta}}{\partial u^{\alpha}}\right|_0 v^{\alpha} \tag{1.55}$$

That is, the matrix of the linear transformation φ'_{x_0} is the Jacobi matrix of the system (1.52). Therefore, in the particular case when M and M' are open subsets of euclidean spaces, φ'_{x_0} is just the derivative of φ. This remark leads to

DEFINITION 1.17 The linear mapping $\varphi'_{x_0} : T_{x_0} M \to T_{y_0} M'$ defined by Proposition 1.10 is called the differential or derivative of φ at x_0. All such notions as $\operatorname{rank}_{x_0} \varphi$, singular points, immersions, embeddings, diffeomorphism, and so on, which were defined earlier for differentiable mappings $f : U \to V$ between open sets of euclidean spaces in terms of their derivative mappings are naturally extended to mappings $\varphi : M \to M'$ of manifolds if one uses the derivative φ'_{x_0}.

Once again, the approach to the same problems for immersed manifolds is through adequate use of their elementary subsets.

As in the case of the spaces $\mathbb{R}^n$, one can easily show that the derivative of the identical mapping is the identity. The formula

$$(\varphi \circ \psi)'_{x_0} = \varphi'_{\psi(x_0)} \circ \psi'_{x_0}$$

holds for all possible compositions of differentiable mappings. The formula

$$(\varphi^{-1})'_{\varphi(x_0)} = (\varphi'_{x_0})^{-1}$$

holds for (local) diffeomorphisms. Detailed proofs of these facts are left to the reader.

An important class of functions is provided by the set $\mathscr{F}(M)$ of all the differentiable mappings $f : M \to R$, where M is an embedded differentiable manifold of $\mathbb{R}^n$. $\mathscr{F}(M)$ is clearly an algebra over the real field $\mathbb{R}$ under the classical sum and product of functions.

Let us consider a point $x_0 \in U \subseteq M$, where U is an open neighborhood of x_0 in M. Then for every differentiable function $f : U \to \mathbb{R}$ we have the derivative $f'_{x_0} : T_{x_0} M \to \mathbb{R}$ [since, by Proposition 1.7, $T_{f(x_0)}\mathbb{R} = \mathbb{R})]$. f'_{x_0} is usually denoted by df and it is called the differential of f.

Now let us take a tangent vector $\underline{v} \in T_{x_0} M$. Then $\underline{v}$ can be regarded as an operator on $\mathscr{F}(M)$ which sends a function to a real number by the formula

$$\underline{v}f = df(\underline{v}) = \sum_{\alpha=1}^{m} v^{\alpha} \left.\frac{\partial f}{\partial u^{\alpha}}\right|_0 \tag{1.56}$$

where the final expression, in terms of a parameterization, follows by formula (1.55). Then the following properties are obviously valid:

$$\underline{v}(\alpha f_1 + \beta f_2) = \alpha(\underline{v}f_1) + \beta(\underline{v}f_2) \qquad \alpha,\ \beta \in \mathbb{R} \tag{1.57}$$

$$\underline{v}(f \cdot g) = (\underline{v}f) \cdot g(x_0) + f(x_0) \cdot (\underline{v}g) \tag{1.58}$$

Note also that $\underline{v} = \underline{0}$ if and only if $\underline{v}f = 0$ for every function f.

Other important functions are those of the form $\varphi : M \to \mathbb{R}^p$, where M is an m-manifold in $\mathbb{R}^n$. Since the values of φ are vectors in $\mathbb{R}^p$, such a function is called a vector field in $\mathbb{R}^p$ along M. In the particular case $p = n$, the values of φ are vectors in the space $\mathbb{R}^n$ where M is embedded, and the following situations can be considered:

(a) A vector field $\varphi : M \to \mathbb{R}^n$ such that $\varphi(x) \in T_xM$ for every $x \in M$ is called a tangent vector field on M.
(b) A vector field $\varphi : M \to \mathbb{R}^n$ such that $\varphi(x) \in N_xM$ for every $x \in M$ is called a normal vector field on M.

If not otherwise specified, any vector field considered will be differentiable in the sense that the corresponding function $\varphi : M \to \mathbb{R}^p$ is such. Thus if $\Phi : U \to M$ is a parameterization, the functions $\varphi_\alpha : \Phi(U) \to \mathbb{R}^n$ defined by $\varphi_\alpha(x_0) = \underline{x}_\alpha^0$—the vector of the natural basis of Φ at x_0 $(\alpha = 1, \ldots, m)$—define tangent vector fields $\underline{x}_\alpha$ on $\Phi(U)$, and they are differentiable because the vectors $\underline{x}_\alpha^0$ are solutions of the system (1.43). [Since differentiability is a local notion, it suffices to take U such that $\Phi(U)$ has implicit equations.] Every other tangent vector field on $\Phi(U)$ can be expressed by means of the natural basis of Φ in the form

$$\underline{v} = \sum_{\alpha=1}^{m} v^{\alpha} \underline{x}_{\alpha} \tag{1.59}$$

where $v^\alpha : \Phi(U) \to \mathbb{R}$ are functions. Since $\underline{x}_\alpha$ are differentiable fields, $\underline{v}$ is differentiable if and only if the coordinate functions v_α are such.

One can also get local differentiable bases for the normal vector fields on $\Phi(U)$. Namely, such bases are given by $\{\text{grad } F_\sigma\}$, where $F_\sigma = 0$ are the implicit equations of $\Phi(U)$.

Let us recall that tangent vectors can be viewed as operators on functions. This yields an action of tangent vector fields on $\mathcal{F}(M)$ defined as follows: Let $f \in \mathcal{F}(M)$ and let $\underline{v}$ be a differentiable tangent vector field on M. Then we obtain a new differentiable function $\underline{v}f$ by

$$(\underline{v}f)(x) = \underline{v}_x f = \sum_{\alpha=1}^{m} v^{\alpha}(x) \left.\frac{\partial f}{\partial u^{\alpha}}\right|_x \tag{1.60}$$

Because of (1.57) and (1.58) we get

$$\underline{v}(\alpha f + \beta g) = \alpha(\underline{v}f) + \beta(\underline{v}g) \quad (\alpha, \beta \in \mathbb{R} \tag{1.61}$$

$$\underline{v}(fg) = (\underline{v}f) \cdot g + f \cdot (\underline{v}g) \tag{1.62}$$

Consider two tangent vector fields $\underline{v}$, $\underline{w}$, and put for every $f \in \mathcal{F}(M)$,

$$[\underline{v}, \underline{w}]f = \underline{v}(\underline{w}f) - \underline{w}(\underline{v}f) \tag{1.63}$$

Then, by (1.60) and with obvious notation, we get

$$[\underline{v}, \underline{w}]f = \sum_{\alpha,\beta=1}^{m} \left(v^{\beta} \frac{\partial w^{\alpha}}{\partial u^{\beta}} - w^{\beta} \frac{\partial v^{\alpha}}{\partial u^{\beta}} \right) \frac{\partial f}{\partial u^{\alpha}} \tag{1.64}$$

Since the right-hand side of (1.64) equals that of (1.63), it does not depend on the parameterization used. On the other hand, if we change the parameterization by

$$u'^{\beta} = u'^{\beta}(u^{\alpha})$$

we have

$$\frac{\partial f}{\partial u'^{\beta}} = \sum_{\alpha=1}^{m} \frac{\partial f}{\partial u^{\alpha}} \frac{\partial u^{\alpha}}{\partial u'^{\beta}} \tag{1.65}$$

and, by definition of $\underline{x}_{\alpha}$,

$$\underline{x}'_{\beta} = \sum_{\alpha=1}^{m} \underline{x}_{\alpha} \frac{\partial u^{\alpha}}{\partial u'^{\beta}} \tag{1.66}$$

Now we can see that the formula

$$[\underline{v}, \underline{w}] = \left(\sum_{\alpha,\beta=1}^{m} v^{\beta} \frac{\partial w^{\alpha}}{\partial u^{\beta}} - w^{\beta} \frac{\partial v^{\alpha}}{\partial u^{\beta}} \right) \underline{x}_{\alpha} \tag{1.67}$$

defines a differentiable tangent vector field on M. Indeed, since (1.65) and (1.66) are identical transformation formulas, it follows that the right-hand side of (1.67) behaves like that of (1.64); that is, it does not depend on the parameterization.

This defines an important operation that associates with every pair $\underline{v}$, $\underline{w}$ a new vector field $[\underline{v}, \underline{w}]$, called the <u>bracket</u> of $\underline{v}$ and $\underline{w}$. Using either (1.63) or (1.67), we get the following properties of this new operation:

$$[\underline{v}, \underline{w}] = -[\underline{w}, \underline{v}] \quad \text{antisymmetry} \tag{1.68}$$

$$[\underline{u}, [\underline{v}, \underline{w}]] + [\underline{v}, [\underline{w}, \underline{u}]] + [\underline{w}, [\underline{u}, \underline{v}]] = 0 \quad \text{the Jacobi identity} \tag{1.69}$$

$$[\alpha \underline{v}_1 + \beta \underline{v}_2, \underline{w}] = \alpha[\underline{v}_1, \underline{w}] + \beta[\underline{v}_2, \underline{w}] \quad \alpha, \beta \in \mathbb{R} \tag{1.70}$$

$$[\underline{v}, f\underline{w}] = f[\underline{v}, \underline{w}] + (\underline{v}f)\underline{w} \tag{1.71}$$

The reader is asked to prove these properties in detail.

Another important notion related to tangent vector fields is provided by

DEFINITION 1.18 Let $\underline{v}$ be a differentiable tangent vector field on M. A path $f:(a,b) \to M$ is called an <u>integral path</u> of $\underline{v}$ if for every $t \in (a,b)$, $\underline{v}_{f(t)}$ is the velocity vector of f at t.

THEOREM 1.4 For every differentiable tangent vector field $\underline{v}$ of M, and for every point $x_0 \in M$, there is some real number $\epsilon > 0$, and a unique path $f:(-\epsilon,+\epsilon) \to M$, which is an integral path of $\underline{v}$ and is such that $f(0) = x_0$.

Proof: Using a parameterization at x_0, we have to look for the local analytic representation of f, $u^\alpha = u^\alpha(t)$, such that

$$\frac{du^\alpha}{dt} = v^\alpha(u^\beta) \qquad u^\alpha(0) = u^\alpha(x_0) \tag{1.72}$$

Hence Theorem 1.4 follows from a classical theorem on ordinary differential equations and the path required is obtained by integrating (1.72).

Let us come back to an arbitrary differentiable vector field $\varphi : M \to \mathbb{R}^p$, $M \subseteq \mathbb{R}^n$. Then, for every $x_0 \in M$, we have the derivative mapping $\varphi'_{x_0} : T_{x_0} \to \mathbb{R}^p$.

If x^λ are coordinates in $\mathbb{R}^p$ and u^α are local coordinates on M at x_0, we have an analytic representation of φ by

$$x^\lambda = x^\lambda(u^\alpha) \tag{1.73}$$

that is, φ is simply an ordered set of p differentiable functions on M. By formula (1.55) this yields

$$\varphi'_{x_0}(\underline{v}) = \left(\sum_{\alpha=1}^{m} \frac{\partial x^\lambda}{\partial u^\alpha} v^\alpha \right)_{\lambda=1,\ldots,p} = [v(x^\lambda)]_{\lambda=1,\ldots,p'} \tag{1.74}$$

where, on the right, we have the action of $\underline{v}$ on x^λ.

We shall denote

$$\underline{v}(\varphi) = \varphi'_{x_0}(\underline{v}) \tag{1.75}$$

and we shall view this formula as defining an action of tangent vectors on general differentiable vector fields.

Following are some immediate, but important, properties of this operation:

$$\underline{v}(\alpha\varphi + \beta\psi) = \alpha\underline{v}(\varphi) + \beta\underline{v}(\psi) \tag{1.76}$$

$$(\alpha\underline{v}_1 + \beta\underline{v}_2)(\varphi) = \alpha\underline{v}_1(\varphi) + \beta\underline{v}_2(\varphi) \qquad \alpha,\beta \in \mathbb{R} \tag{1.77}$$

$$\underline{v}(f\varphi) = (\underline{v}f)\varphi(x_0) + f(x_0)\underline{v}(\varphi) \qquad f: M \to \mathbb{R} \tag{1.78}$$

$$\underline{v}(\varphi \cdot \psi) = \underline{v}(\varphi) \cdot \psi(x_0) + \varphi(x_0) \cdot \underline{v}(\psi) \tag{1.79}$$

where the dot denotes scalar product of vectors in $\mathbb{R}^p$. The proofs are left to the reader as an exercise.

We note that the operation (1.75) can be defined for tangent vector fields $\underline{v}$ as well. We simply let x_0 vary in M, and the result will be a new, general vector field $\underline{v}(\varphi)$.

For example, if φ is also a tangent field, an easy computation yields

$$\underline{v}(\varphi) - \varphi(\underline{v}) = [\underline{v}, \varphi] \tag{1.80}$$

and if we have three tangent vector fields, $\underline{u}$, $\underline{v}$, $\underline{w}$, we get

$$[\underline{u}, \underline{v}](\underline{w}) = \underline{u}(\underline{v}(\underline{w})) - \underline{v}(\underline{u}(\underline{w})) \tag{1.81}$$

As usual, we end this section by noting that the case of the immersed manifolds has to be studied by using all the corresponding elementary subsets. We ask the reader to verify the euclidean character of the results described above.

EXERCISES

1.35 Consider the mapping of unit spheres $f: S^m \to S^n$ $(m \leq n)$ defined by

$$f(x^1, \ldots, x^{m+1}) = (1, \ldots, 0, x^1, \ldots, x^{m+1})$$

Prove that f is a differentiable mapping and, moreover, that f is an embedding of S^m in S^n.

1.36 Define $h: S^n \to S^n$ by $h(x) = -x$ $(x \in S^n)$. Prove that h is a diffeomorphism.

1.37 Consider the product $M \times N$ of two differentiable manifolds, as defined in Exercise 1.18. Prove that the two projections $p: M \times N \to M$ and $q: M \times N \to N$ of the cartesian product onto its factors are differentiable mappings. Write the equations of the differentials p'_x, q'_x $(x \in M \times N)$, and compute their ranks.

1.38 Prove that the orthogonal projection p of the sphere $S^2 \subset R^3$ onto the plane $x^3 = 0$ is a differentiable mapping. Write the equations of p'_x, and study its rank when x runs through S^2.

1.39 Since the circle S^1 is embedded in $\mathbb{R}^2$, we can define (by Exercise 1.18) $S^1 \times S^1$ as an embedded manifold in $\mathbb{R}^4$. Prove that one can construct an embedding of $S^1 \times S^1$ in $\mathbb{R}^3$ as well. [Hint: $S^1 \times S^1$ is diffeomorphic to a torus.]

1.40 Write the equations of the differential of the embedding of $S^1 \times S^1$ in $\mathbb{R}^3$ which you have constructed in Exercise 1.39.

1.41 Consider the sphere S^2 in $\mathbb{R}^3$ and the point $n = (0, 0, 1)$. Denote by C_x the circle of center 0 passing through n and x. Furthermore, denote by $\ell(x)$ the length of the shortest arc nx of C_x. Study the differentiability of the function $\ell : S^2 \to \mathbb{R}$ thereby obtained. Assume that $x_0 \in S^2$, $x_0 \neq (0, 0, \pm 1)$, and compute $\underline{v}\ell$, where $\underline{v}$ is the unit tangent vector of C_{x_0} at x_0 which points toward n.

1.42 Show that the vectors $\underline{v}$ of Exercise 1.41 define a differentiable tangent vector field on $S^2 \backslash \{(0, 0, \pm 1)\}$, but that this field cannot be extended to a differentiable tangent vector field on the whole sphere S^2.

1.43 Prove that there are three vector fields tangent to the unit sphere $S^3 \subset \mathbb{R}^4$ which are linearly independent at every point of S^3. [Hint: The fields are defined in $\mathbb{R}^4$ by $(x^2, -x^1, x^4, -x^3)$ and similar points, where $(x^1, x^2, x^3, x^4) \in S^3$.]

1.44 Compute the bracket of every pair of the vector fields obtained in Exercise 1.43.

1.45 Define on $\mathbb{R}^2$ the vector fields

$$\underline{u} = x^1 \frac{\partial}{\partial x^1} - x^2 \frac{\partial}{\partial x^2} \qquad \underline{v} = x^2 \frac{\partial}{\partial x^1} \qquad \underline{w} = x^1 \frac{\partial}{\partial x^2}$$

Prove that the brackets $[\underline{u}, \underline{v}]$, $[\underline{v}, \underline{w}]$, $[\underline{w}, \underline{u}]$ are linear combinations of the fields $\underline{u}$, $\underline{v}$, $\underline{w}$.

1.46 Find the integral paths of the vector fields $\underline{u}$, $\underline{v}$, $\underline{w}$ of Exercise 1.45.

1.7 ORIENTABLE AND NONORIENTABLE MANIFOLDS

Another important type of property, which is of a topological nature and can be discussed via the tangent spaces, is that of <u>orientability</u>. Logically, orientation is similar to the distinction between "left" and "right."

Let us begin by considering an n-dimensional real vector space L. The set B of all the bases of L can be uniquely divided (up to notation) into a disjoint union of two nonempty subsets

$$B = \overset{+}{B} \cup \overset{-}{B} \tag{1.82}$$

such that if either $b_1, b_2 \in \overset{+}{B}$ or $b_1, b_2 \in \overset{-}{B}$, then $b_2 = tb_1$, where t is an $n \times n$ matrix with positive determinant ($\det t > 0$), and if $b_1 \in B^+$, $b_2 \in B^-$, then $b_2 = tb_1$ with $\det t < 0$. Indeed, it suffices to take an arbitrary $b \in B$, and to put $b_1 \in B^+$ iff $b_1 = tb$ with $\det t > 0$ and $b_2 \in B^-$ iff $b_2 = tb$ with $\det t < 0$.

This property is called the <u>orientability</u> of L and an <u>orientation</u> of L consists in a choice of one of the subsets above (e.g., B^+) as the set of <u>positively oriented bases</u>.

The orientability of L is equivalent to the existence of a function

$$\epsilon : B \to \{+1, -1\} \tag{1.83}$$

which has the property

$$\epsilon(b_2) = \text{sign det } t \cdot \epsilon(b_1) \tag{1.84}$$

if $b_2 = tb_1$.

Recall that

$$\text{sign } \alpha = \begin{cases} +1 & \text{if } \alpha > 0 \\ -1 & \text{if } \alpha < 0 \end{cases}$$

Indeed, if we have the decomposition (1.82), and if we define $\epsilon | B^+ = +1$, $\epsilon | B^- = -1$, (1.84) is satisfied. Conversely, if ϵ exists, and if we define $B^+ = \{b | b \in B, \epsilon(b) = +1\}$, $B^- = \{b | b \in B, \epsilon(b) = -1\}$, we obtain (1.82), and (1.84) assures the validity of the conditions stated for such a decomposition. Furthermore, an orientation of L is a choice of such a function ϵ.

Now consider an embedded m-dimensional manifold $M \subseteq \mathbb{R}^n$. For future utilization, we recall first that a subset $U \subseteq M$ is said to be connected if any two points of U can be joined by a path in U. The maximally connected subsets of M are called the connected components of M.

The tangent spaces T_xM $(x \in M)$ are orientable, and we can choose an orientation ϵ_x in each one of them. Since the values of ϵ_x are ± 1, it is natural to say that the orientations ϵ_x depend continuously on x if, for every open connected subset $U \subseteq M$ and for every system of m continuous vector fields $\underline{v}_\alpha$ $(\alpha = 1, \ldots, m)$ over U which are linearly independent at every point $x \in U$ [and thereby define bases b(x) at every such point x], $\epsilon_x(b(x)) =$ const. If the ϵ_x satisfy this condition, we say that we have a continuous field of orientations on M. Now we can state

DEFINITION 1.19 An embedded manifold $M \subseteq \mathbb{R}^n$ is said to be orientable if it has a continuous field of orientations, and every such field defines an orientation of M. If such a field does not exist, M is said to be nonorientable.

Orientability can also be characterized in several other equivalent ways, and we shall consider some of these.

THEOREM 1.5 The manifold M embedded in $\mathbb{R}^n$ is orientable if and only if it has an atlas $\mathscr{A}$, defined by the parameterizations $\Phi_\tau : U_\tau \to \mathbb{R}^n$ (U_τ are open subsets in $\mathbb{R}^m$), such that for every pair Φ_τ $\Phi_{\tau'}$, the transition functions $u'^\beta(u^\alpha)$ have a positive jacobian at every point.

Proof: Suppose that the atlas $\mathscr{A}$ exists. Take $x \in M$ and some Φ_τ such that $x \in$ range Φ_τ. Next, define ϵ_x by the condition that it is +1 for the natural basis of Φ_τ at x. This completely defines ϵ_x, and ϵ_x does not depend on the choice of Φ_τ because of (1.66) and of the hypothesis of positive jacobians. Then the ϵ_x define a continuous field of orientations on M. Indeed, locally, vector fields $\underline{v}_\alpha$ as in the definition of the continuity of ϵ_x can be expressed as

$$\underline{v}_\alpha = \sum_{\alpha=1}^{m} v_\alpha^\beta \underline{x}_\beta \qquad (1.85)$$

where $\underline{x}_\beta$ is the natural basis of some Φ_τ. This yields for the basis b(x) defined by these vectors,

$$\epsilon(b(x)) = \text{sign}\det(v_\alpha^\beta)\cdot\epsilon(\{\underline{x}_\beta\}) = \text{const}$$

Now let us prove the converse. Suppose that ϵ_x is a continuous field of orientations on M, and let $\mathscr{A}' = \{\Phi_\theta\}$ be an atlas on M such that the ranges of all of its parameterizations are connected. (The existence of such an atlas is obvious.) Then, because of the continuity, ϵ_x is constant for the natural bases of every parameterization of $\mathscr{A}'$.

Now, starting with $\mathscr{A}'$ we can construct a new atlas $\mathscr{A}$ as follows. Every parameterization of $\mathscr{A}'$ such that $\epsilon_x = +1$ for its natural bases is included in $\mathscr{A}$. If $\Phi_{\theta'}$ is a parameterization of $\mathscr{A}'$ such that for its natural bases $\epsilon_x = -1$, then we change the sign of one of the parameters and include in $\mathscr{A}$ the new parameterization thus obtained. Obviously, $\mathscr{A}$ is the atlas required by Theorem 1.5, which is thereby proven.

Next, let us consider the normal space N_xM of M at x, and let us denote by η_x its orientation functions. Let us give the space $\mathbb{R}^n$ the usual orientation defined by

$$\epsilon(\{\underline{e}_i\}) = 1$$

where $\underline{e}_i$ is the canonical basis $\underline{e}_i(\delta_i^j)$ of $\mathbb{R}^n$ ($\delta_i^i = 1$, $\delta_i^j = 0$ for $i \neq j$). Then every orientation ϵ_x on T_xM defines a unique orientation η_x on N_xM by

$$\eta_x(b') = \epsilon(b \cup b') \qquad (1.86)$$

where b' is a basis of N_xM and b is an arbitrary positively oriented basis of T_xM. In a similar manner, every η_x defines an ϵ_x.

Now one can define the notion of a <u>continuous field of normal orientations</u> η_x in the same manner as we defined a continuous field ϵ_x. Moreover, if ϵ_x is a continuous field, then the field η_x defined by (1.86) is continuous, and conversely. Thereby, we have proved

PROPOSITION 1.11 The manifold M embedded in $\mathbb{R}^n$ is orientable if and only if it has a continuous field of normal orientations.

COROLLARY 1.2 An embedded hypersurface M of $\mathbb{R}^n$ is orientable if and only if it has a continuous field of unit normal vectors.

Proof: If $\underline{n}(x)$ is such a field, then $\eta_x(n(x)) = 1$ defines on M a continuous field of normal orientations, and M is orientable. Conversely, let M be orientable and let η_x be a continuous field of normal orientations on M. Then, at every point $x \in M$ there is a unique unit positively oriented normal vector, and this defines a continuous field of unit normal vectors on M since it is given locally by

$$\eta\left(\frac{\text{grad } F}{|\text{grad } F|}\right) \frac{\text{grad } F}{|\text{grad } F|}$$

where $F(x^i) = 0$ is a local implicit equation of M.

We can now give examples of orientable and nonorientable manifolds. First we have

PROPOSITION 1.12 Every elementary m-dimensional subset of $\mathbb{R}^n$ is an orientable manifold.

Proof: Let us assume that we have such a subset Γ, defined by the implicit equations $F_\sigma(x^i) = 0$. Then grad F_σ ($\sigma = 1, \ldots, n - m$) define normal bases b(x) for Γ, and if we put $\eta_x(b(x)) = 1$, we get a continuous field of normal orientations. Q.E.D.

As a particular case of Proposition 1.12 we have

COROLLARY 1.3 Every hypersurface of $\mathbb{R}^n$ that admits a global implicit equation is orientable.

The converse of Corollary 1.3—that every orientable hypersurface embedded in $\mathbb{R}^n$ admits a global implicit equation—is also true but much harder to prove. Moreover, one can prove that every topologically closed embedded hypersurface of $\mathbb{R}^n$ has a global implicit equation, and is therefore orientable. We have thereby a large class of orientable manifolds, and we point out a single example: the hypersphere S^{n-1}.

The typical example of a nonorientable manifold is the <u>Möbius band</u>, which we shall now describe (see Fig. 1.5). Let us consider the open segment $AB \parallel Ox^3$ through the point C situated on Ox^1 such that $OC = 2$. Moreover, let C be the midpoint of AB and $AB = 2$. We assume that this segment is moved by a motion composed of (a) a rotation of point C about Ox^3; and (b) a simultaneous rotation of AB about C in plane COx^3, such that when C rotates through angle ω, AB rotates about C through $\omega/2$. Thus, after a full rotation of C, segment AB is sent to BA.

The set of points in $\mathbb{R}^3$ that is the union of all the positions of AB during a full rotation of C is called a <u>Möbius band</u>. This set can be also conceived

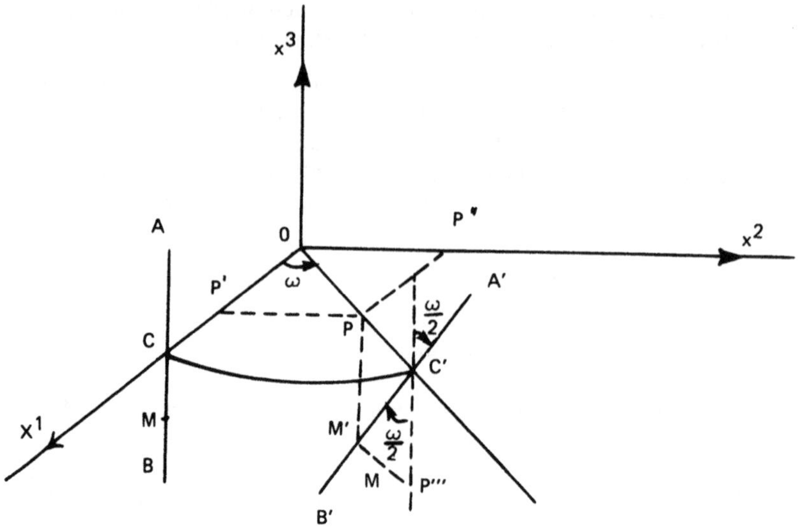

Figure 1.5

as the result of gluing up edges AB and CD after a twist in rectangle ABCD of Fig. 1.6.

We shall show that the Möbius band is a nonorientable surface embedded in $\mathbb{R}^3$. To do this, we consider $M \in AB$ such that $CM = u$, and let M' be the position of M after the rotation of C through the angle ω. From Fig. 1.5 we get the coordinates of M':

$$
\begin{aligned}
x^1 &= \left(2 - u \sin\frac{\omega}{2}\right) \cos \omega \\
x^2 &= \left(2 - u \sin\frac{\omega}{2}\right) \sin \omega \qquad (1.87) \\
x^3 &= u \cos \frac{\omega}{2}
\end{aligned}
$$

Equations (1.87) can be thought of as the "parametric equations" of the Möbius band, but they do not define a global parameterization in our usual sense, because we must take, for instance,

$-1 < u < 1 \quad 0 \leq \omega < 2\pi$

which is not an open subset of the plane $\{(u,\omega)\}$.

If we take in (1.87)

$$-1 < u < 1 \qquad \omega_0 < \omega < \omega_0 + 2\pi \tag{1.88}$$

the range of the map (1.87) is the Möbius band without the position of segment AB, which corresponds to rotation angle ω_0. From equations (1.87) and Fig. 1.5 it follows that (1.87) defines a diffeomorphism of the domain (1.88). Hence the domain B_0 defined by the band minus one segment $A_0 B_0$ is an embedded manifold in $\mathbb{R}^3$, and therefore every point of B_0 has a neighborhood U in $\mathbb{R}^3$ such that $U \cap B_0$ is an elementary set. Since every point of the Möbius band is interior to such a domain B_0, this neighborhood property holds for the whole band, which is thereby an embedded surface in $\mathbb{R}^3$.

Moreover, for every $\omega_0 \in [0, 2\pi)$ the restriction of (1.87) to (1.88) is a parameterization whose range is the band minus a position of AB. Therefore, by choosing two distinct values of ω_0, for instance, $\omega_0 = 0$ and $\omega_0 = \pi$, we get two parameterizations which make up an atlas.

Now, if the Möbius band were orientable, we should have a continuous normal unit vector field $\underline{N}$ on it (Corollary 1.2). We know that the natural basis of the tangent space consists of the vectors

$$\underline{a}\left(\frac{\partial x^1}{\partial u}, \frac{\partial x^2}{\partial u}, \frac{\partial x^3}{\partial u}\right) \qquad \underline{b}\left(\frac{\partial x^1}{\partial \omega}, \frac{\partial x^2}{\partial \omega}, \frac{\partial x^3}{\partial \omega}\right)$$

whence obviously

$$\underline{N} = \pm \frac{\underline{a} \times \underline{b}}{|\underline{a} \times \underline{b}|} \tag{1.89}$$

where $\times$ denotes a vector product in three-dimensional space.

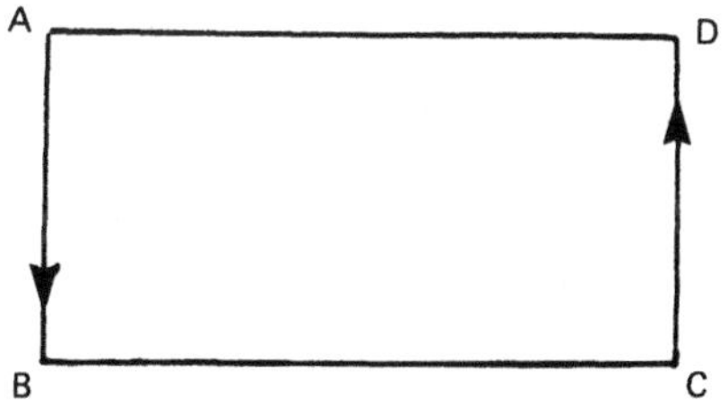

Figure 1.6

Since the two parameterizations (1.87) and (1.88) with $\omega_0 = 0$, $\omega_0 = \pi$ which cover the band have ranges with a nonempty intersection, we must have for both the same choice of the sign in (1.89). Otherwise, the continuity of $\underline{N}$ would be contradicted. Let us assume that we choose the sign +.

Then the computation yields that for $u = 0$ and $\omega = 2\pi$ (which belongs to the domain of the first parameterization above, $\underline{N} = (1, 0, 0)$, and for $u = 0$ and $\omega = 0$ (which belongs to the domain of the second parameterization), $\underline{N} = (-1, 0, 0)$. This contradicts the uniqueness of the vector of the field $\underline{N}$ at C.

The contradiction shows that the Möbius band cannot be orientable. Intuitively, if we start with a normal unit vector at C of Fig. 1.6 and move it continuously along the curve described by C, we come back to C with a reversed orientation.

We end this section with a few other remarks.

First, it is easy to count how many orientations are on an orientable manifold. Indeed, if we assume that the manifold is connected, then we see, by "propagating the orientation function along paths," that it admits exactly two orientations. If the manifold has several connected components, we may combine orientation functions on these components in all possible ways.

Second, Theorem 1.5 shows that, as a matter of fact, orientability is an "internal" property and does not depend on the ambient space. Hence it is possible to consider orientability for immersed manifolds by using atlases. This also shows that we can speak of orientability and orientation in any euclidean space.

EXERCISES

1.47 Assume that M and N are two orientable differentiable manifolds. Prove that the cartesian product $M \times N$ as defined in Exercise 1.18 is an orientable manifold as well.

1.48 Let M and N be two oriented differentiable manifolds, and let ϵ and θ be the continuous fields of orientations on M and N, respectively. Assume that $f: M \to N$ is a diffeomorphism. f is said to be <u>orientation preserving</u> if $\theta(f'_x(b_x)) = \epsilon(b_x)$ for every point $x \in M$ and every basis b_x of T_xM. Prove that f is orientation preserving iff the jacobian of the coordinate expression of f with respect to parameterizations of M, N at x and f(x), which belong to the oriented atlases of M, N, respectively, is everywhere positive. Prove that this property does not depend on the choice of the parameterizations above.

1.49 Give a detailed definition of an orientation of the unit sphere S^n, and find a corresponding oriented atlas. Consider the diffeomorphism $h: S^n \to S^n$ defined in Exercise 1.36 and prove that h preserves the orientation iff n is odd. [Hint: Use the normal to the sphere to establish a relation between the orientation of S^n and the orientation of the ambient space $\mathbb{R}^{n+1}$.]

1.50 Consider the projective space P^n with the atlas defined in Exercise 1.24. Prove that this atlas is oriented if n is even and is not oriented if n is odd.

1.8 TENSORS AND TENSOR FIELDS ON MANIFOLDS

Another consequence of the existence of the tangent space is the possibility of defining, by algebraic means, many related linear spaces, which leads to a computational algorithm for invariants on manifolds.

Take $x \in M \subseteq \mathbb{R}^n$, where M is an embedded m-dimensional differentiable manifold, and let T_xM be the tangent space at x. Then, for every parameterization $\Phi : U \to \mathbb{R}^n$ of M at x, which is defined by the equations

$$\underline{x} = \underline{x}(u^\alpha) \quad \alpha = 1, \ldots, m \tag{1.90}$$

we have the corresponding natural basis of T_xM given by the vectors

$$\underline{x}_\alpha = \frac{\partial \underline{x}}{\partial u^\alpha} \tag{1.91}$$

and for $\underline{v} \in T_xM$ we have

$$\underline{v} = \sum_{\alpha=1}^{m} v^\alpha \underline{x}_\alpha \tag{1.92}$$

v^α being the internal coordinates of $\underline{v}$.

Since symbols of derivatives and sums, like those in (1.91) and (1.92), will arise many times in later formulas, we make the following important notation conventions:

(a) The partial derivative with respect to a variable u^α will be indicated by the index α, for example,

$$\varphi_{\alpha\beta\gamma} = \frac{\partial^3 \varphi}{\partial u^\alpha \partial u^\beta \partial u^\gamma}$$

(b) Every monomial containing indexed variables, where one of the upper indices α appears also as a lower index, denotes a sum over α, where α runs through all its possible values. For example,

$$v^\alpha \underline{x}_\alpha = \sum_{\alpha=1}^{m} v^\alpha \underline{x}_\alpha \qquad t^i_j \xi_i \eta^j = \sum_{i,j=1}^{n} t^i_j \xi_i \eta^j$$

For a correct use of convention (b), we must carefully preserve the place of the indices and always use different summation indices. For example, the expression

$$t^{\alpha}_{\beta\alpha} t^{\alpha}_{\sigma\alpha}$$

is meaningless (except when a special meaning is indicated for it).

Convention (b) is called the Einstein convention. The repeating indices are called summation indices, and the others are free indices. (For example, in the expression $a_{ij}\xi^j$, j is a summation index, and i is a free index.)

For further applications we also introduce

DEFINITION 1.20 An element $\underline{v} \in T_xM$ is also called a contravariant vector of M at x and v^{α} of (1.92) are called its components.

From formula (1.66) we know that, by a change

$$u'^{\beta} = u'^{\beta}(u^{\alpha}) \tag{1.93}$$

of the parameterization used, the natural bases change by the rule

$$\underline{x}'_{\beta} = \frac{\partial u^{\alpha}}{\partial u'^{\beta}} \underline{x}_{\alpha} \tag{1.94}$$

Hence, for a given vector $\underline{v}$, we have

$$\underline{v} = v^{\alpha}\underline{x}_{\alpha} = v'^{\beta}\underline{x}'_{\beta} = v'^{\beta} \frac{\partial u^{\alpha}}{\partial u'^{\beta}} \underline{x}_{\alpha} \tag{1.95}$$

and we see that the transformation rule for its components is

$$v^{\alpha} = \frac{\partial u^{\alpha}}{\partial u'^{\beta}} v'^{\beta} \tag{1.96}$$

Note that for every parameterization we have $\partial u^{\gamma}/\partial u^{\alpha} = \delta^{\gamma}_{\alpha}$, where

$$\delta^{\gamma}_{\alpha} = \begin{cases} 0 & \text{for } \alpha \neq \gamma \\ 1 & \text{for } \alpha = \gamma \end{cases} \tag{1.97}$$

This symbol (already encountered by us) is called the Kronecker delta. (All the Greek indices run from 1 to m.)

Therefore, by using the chain rule for partial derivatives we obtain

$$\frac{\partial u^{\gamma}}{\partial u'^{\beta}} \frac{\partial u'^{\beta}}{\partial u^{\alpha}} = \delta^{\gamma}_{\alpha} \tag{1.98}$$

In view of (1.98), (1.94) yields

$$\frac{\partial u'^{\beta}}{\partial u^{\gamma}} \underline{x}'_{\beta} = \frac{\partial u'^{\beta}}{\partial u^{\gamma}} \frac{\partial u^{\alpha}}{\partial u'^{\beta}} \underline{x}_{\alpha} = \delta^{\alpha}_{\gamma} \underline{x}_{\alpha} = \underline{x}_{\gamma}$$

that is,

$$\underline{x}_{\gamma} = \frac{\partial u'^{\beta}}{\partial u^{\gamma}} \underline{x}'_{\beta} \tag{1.99}$$

and similarly, (1.96) yields

$$v'^{\beta} = \frac{\partial u'^{\beta}}{\partial u^{\gamma}} v^{\gamma} \tag{1.100}$$

The last formula is called the transformation rule for the components of a contravariant vector, and (1.95) shows that whenever (1.100) holds, the vector $\underline{v}$ is well defined. That is, (1.100) is the criterion for recognizing the components of a contravariant vector.

Now, we shall proceed to related linear spaces.

First, there is the dual space $T^*_x M$, which is the space of all the linear maps $\omega : T_x M \to \mathbb{R}$, and we give

DEFINITION 1.21 Every $\omega \in T^*_x M$ is called a covariant vector, or simply a covector, of M at x.

For example, we see from Sec. 1.6 that for every differentiable function $f : U \to \mathbb{R}$, where U is an open neighborhood of x in M, its differential df(x) at x, is such a covariant vector. As a matter of fact, it can be shown that these are the only possible covariant vectors.

In particular, the differentials du^{α} of the local parameters belong to $T^*_x M$, and for any $\underline{v} \in T_x M$ we have

$$du^{\alpha}(\underline{v}) = \underline{v}(u^{\alpha}) = v^{\beta} \frac{\partial u^{\alpha}}{\partial u^{\beta}} = v^{\alpha} \tag{1.101}$$

Then for an arbitrary $\omega \in T^*_x M$ we have

$$\omega(\underline{v}) = v^{\beta} \omega(\underline{x}_{\beta}) = \omega_{\beta}\, du^{\beta}(v) \tag{1.102}$$

where

$$\omega_{\beta} = \omega(\underline{x}_{\beta}) \tag{1.103}$$

On the other hand, we have by (1.101),

$$du^{\alpha}(\underline{x}_{\beta}) = \delta^{\alpha}_{\beta} \tag{1.104}$$

whence the du^{α} are easily seen to be linearly independent. Then we see by (1.102) that du^{α} is a basis in T^*_xM for every x in the range of the parameterization $(\Phi, (u^{\alpha}))$. This will be called the natural cobasis of the parameterization Φ (at x, of course), and (1.103) are called the components of ω.

A change of the parameterization yields, by means of (1.99),

$$du'^{\alpha}(\underline{x}_{\gamma}) = \frac{\partial u'^{\alpha}}{\partial u^{\gamma}}$$

whence we deduce the following transformation law of the natural cobases:

$$du'^{\alpha} = \frac{\partial u'^{\alpha}}{\partial u^{\gamma}} du^{\gamma} \tag{1.105}$$

Similarly, from (1.94) and (1.103) we get

$$\omega'_{\beta} = \frac{\partial u^{\alpha}}{\partial u'^{\beta}} \omega_{\alpha} \tag{1.106}$$

which is the transformation law for the components of a covector, and yields the criterion to recognize the components of a covector. Let us also note the following immediate consequence of (1.102):

$$\omega(\underline{v}) = \omega_{\alpha} v^{\alpha} \tag{1.107}$$

Furthermore, we can consider the following general case:

DEFINITION 1.22 Consider a function

$$t : T_xM \times T_xM \times T^*_xM \times \cdots \times T^*_xM \times T_xM \to \mathbb{R} \tag{1.108}$$

where the total number of the factors T_xM is q, the total number of factors T^*_xM is p, and the order of the factors is well defined. If this function is linear with respect to all its arguments, we say that t is a p-times contravariant and q-times covariant tensor of M at x, or a tensor of type (p, q).

REMARK It is very important to note that there are many kinds of tensors of type (p, q), each of them being defined by the corresponding order of the factors of the cartesian product (1.108). Whenever we speak of the set of (p, q) tensors at x, we have in mind only one of these kinds of tensors.

Let us denote by $\mathcal{T}^p_q(M;x)$ the set of the (p, q) tensors at $x \in M$ (of a given kind!), and consider also $\mathcal{T}^0_0 = \mathbb{R}$.

From Definition 1.21, it follows that $\mathcal{T}_1^0(M;x) = T_x^*M$. Moreover, by (1.107) every vector $\underline{v} \in T_xM$ becomes a linear function $\underline{v}: T_x^*M \to \mathbb{R}$ if we set $\underline{v}(\omega) = \omega(\underline{v})$ and conversely. Hence we have $\mathcal{T}_0^1(M;x) = T_xM$, and we see that the notion of a tensor is general, and contains the scalars, vectors, and covectors as particular cases.

By means of a parameterization at x, the values of the (p, q) tensor t of (1.108) can be calculated as follows:

$$t(\underline{v}_1, \underline{v}_2, \omega^1, \ldots, \omega^p, \underline{v}_q)$$

$$= t\left(v_1^{\alpha_1} \underline{x}_{\alpha_1}, v_2^{\alpha_2} \underline{x}_{\alpha_2}, \omega^1_{\beta_1} du^{\beta_1}, \ldots, \omega^p_{\beta_p} du^{\beta_p}, v_q^{\alpha_q} \underline{x}_{\alpha_q}\right)$$

$$= t_{\alpha_1 \alpha_2}{}^{\beta_1 \cdots \beta_p}{}_{\cdots \alpha_q} v_1^{\alpha_1} v_2^{\alpha_2} \omega^1_{\beta_1} \cdots \omega^p_{\beta_p} v_q^{\alpha_q}$$

where we have set

$$t_{\alpha_1 \alpha_2}{}^{\beta_1 \cdots \beta_p}{}_{\cdots \alpha_q} = t(\underline{x}_{\alpha_1}, \underline{x}_{\alpha_2}, du^{\beta_1}, \ldots, du^{\beta_p}, \underline{x}_{\alpha_q}) \tag{1.109}$$

Note that the free places above the α and those under the β are very important since they indicate the order of the vectorial and covectorial arguments of t. However, if there is no chance of confusion, we can write simply

$$t^{\beta_1 \cdots \beta_p}_{\alpha_1 \cdots \alpha_q}$$

The $t^{\beta_1 \cdots \beta_p}_{\alpha_1 \cdots \alpha_q}$ defined by (1.109) are called the <u>components</u> of the tensor t with respect to the given parameterization. The components with respect to a second parameterization are, in view of (1.94) and (1.105),

$$t'^{\mu_1 \cdots \mu_p}_{\lambda_1 \cdots \lambda_q} = t\left(\underline{x}'_{\lambda_1}, \underline{x}'_{\lambda_2}, du'^{\mu_1}, \ldots, du'^{\mu_p}, \underline{x}'_{\lambda_q}\right)$$

$$= \frac{\partial u'^{\mu_1}}{\partial u^{\beta_1}} \cdots \frac{\partial u'^{\mu_p}}{\partial u^{\beta_p}} \frac{\partial u^{\alpha_1}}{\partial u'^{\lambda_1}} \cdots \frac{\partial u^{\alpha_q}}{\partial u'^{\lambda_q}} \cdot t^{\beta_1 \cdots \beta_p}_{\alpha_1 \cdots \alpha_q} \tag{1.110}$$

(1.110) is the transformation law of the components of a (p, q) tensor and yields the criterion for recognizing such components.

At this point we shall introduce several important operations on tensors. First, since linear functions can be added and multiplied by a scalar, we see that $\mathcal{T}^p_q(M;x)$ is a real linear space. Clearly, if $t, \tilde{t} \in \mathcal{T}^p_q(M;x)$, the components of $t + \tilde{t}$ and λt are, respectively,

$$t^{\beta_1\cdots\beta_p}_{\alpha_1\cdots\alpha_q} + \tilde{t}^{\beta_1\cdots\beta_p}_{\alpha_1\cdots\alpha_q}, \quad \lambda t^{\beta_1\cdots\beta_p}_{\alpha_1\cdots\alpha_q}$$

Next, let $t \in \mathcal{T}^p_q(M;x)$ and $\tilde{t} \in \mathcal{T}^s_u(M;x)$. We define their <u>tensor product</u> $\bar{t} = t \otimes \tilde{t}$ as the $(p + s,\ q + u)$ tensor given by

$$\bar{t}\,(\underline{v}_1,\ldots,\underline{v}_{p+s},\ \omega^1,\ldots,\omega^{q+u})$$

$$= t(\underline{v}_1,\ldots,\underline{v}_p,\ \omega^1,\ldots,\omega^q)\,\tilde{t}(\underline{v}_{p+1},\ldots,\underline{v}_{p+s},\ \omega^{q+1},\ldots,\omega^{q+u}) \tag{1.112}$$

or, componentwise, by

$$\bar{t}^{\beta_1\cdots\beta_{p+s}}_{\alpha_1\cdots\alpha_{q+u}} = t^{\beta_1\cdots\beta_p}_{\alpha_1\cdots\alpha_q}\tilde{t}^{\beta_{p+1}\cdots\beta_{p+s}}_{\alpha_{q+1}\cdots\alpha_{q+u}} \tag{1.113}$$

This product is associative and distributive, but it may not be commutative.

A last algebraic operation to be considered here is the <u>contraction of indices</u>. Let $t \in \mathcal{T}^q_p(M;x)$ with the components $t^{\beta_1\cdots\beta_q}_{\alpha_1\cdots\alpha_p}$ and $p, q \geq 1$. Consider

$$\tilde{t}^{\beta_1\cdots\beta_{q-1}}_{\alpha_1\cdots\alpha_{p-1}} = t^{\beta_1\cdots\beta_{q-1}\lambda}_{\alpha_1\cdots\alpha_{p-1}\lambda} \tag{1.114}$$

where, of course, in the right-hand side λ is a summation index.

Then a simple computation which uses (1.98) and (1.110) shows that $\tilde{t}^{\cdots}_{\cdots}$ satisfies the criterion (1.110) and defines a tensor $\tilde{t} \in \mathcal{T}^{q-1}_{p-1}(M;x)$. This $\tilde{t}$ is said to be the <u>result of the contraction of β_q and α_p in</u> t. Obviously, in a similar manner, any upper index can be <u>contracted</u> with an arbitrary lower index.

Now, let us provide some important examples. First, the usual scalar product in $\mathbb{R}^n$ defines a bilinear function

$$g : T_xM \times T_xM \to \mathbb{R} \tag{1.115}$$

by $g(\underline{v}, \underline{w}) = \underline{v} \cdot \underline{w}$. We get thereby a tensor $g \in \mathcal{T}_2^0(M;x)$, called the first fundamental tensor or the metric tensor of M at x. It is clear that this tensor is related to metric properties (length of vectors, angle between two vectors, etc.), and one has

$$g(\underline{v}, \underline{w}) = g(\underline{w}, \underline{v}) \tag{1.116}$$

$$g(\underline{v}, \underline{v}) \geq 0 \quad g(\underline{v}, \underline{v}) = 0 \text{ iff } \underline{v} = 0 \tag{1.117}$$

Because of (1.116) we say that this tensor is symmetric, and because of (1.117) that it is positive definite.

The components of g are

$$g_{\alpha\beta} = g(\underline{x}_\alpha, \underline{x}_\beta) = \underline{x}_\alpha \cdot \underline{x}_\beta \tag{1.118}$$

and (1.116) and (1.117) yield $g_{\alpha\beta} = g_{\beta\alpha}$, and the fact that $g_{\alpha\beta}v^\alpha v^\beta$ is a positive definite quadratic form on T_xM, whence $(g_{\alpha\beta})$ is a nondegenerate $m \times m$ matrix, and has a symmetric inverse matrix, which we shall denote by $(g^{\alpha\beta})$ and which is characterized by

$$g_{\alpha\lambda}g^{\lambda\beta} = \delta_\alpha^\beta \tag{1.119}$$

Using g, we can define

$$\flat : T_xM \to T_x^*M \tag{1.120}$$

by

$$(\flat \underline{v})(\underline{w}) = g(\underline{v}, \underline{w}) \quad \underline{w} \in T_xM \tag{1.121}$$

that is, componentwise

$$v_\alpha w^\alpha = g_{\alpha\beta} v^\alpha w^\beta$$

where the v_α (α is a lower index!) denote the components of the covector $\flat\, \underline{v}$, the v^α (α is an upper index!) are the components of the vector $\underline{v}$, and the w^α are the components of an arbitrary vector $\underline{w} \in T_xM$. In other words, (1.121) provides

$$v_\alpha = g_{\alpha\beta} v^\alpha \tag{1.122}$$

In addition, we shall define

$$\sharp : T_x^*M \to T_xM \tag{1.123}$$

by

$$g(\sharp\,\omega, \underline{w}) = \omega(\underline{w}) \quad \underline{w} \in T_xM \tag{1.124}$$

This is a well-defined linear map since, if we denote by ω_α the components of ω and by ω^α those of the required vector $\sharp\omega$, (1.124) yields

$$g_{\alpha\beta}\omega^\alpha w^\beta = \omega_\beta w^\beta \tag{1.125}$$

that is,

$$g_{\alpha\beta}\omega^\alpha = \omega_\beta \tag{1.126}$$

whence, in view of (1.119) we get

$$\omega^\alpha = g^{\alpha\beta}\omega_\beta \tag{1.127}$$

It is obvious that $\flat$ and $\sharp$ are inverse to each other, and therefore, they are isomorphisms of linear spaces. They are called the musical isomorphisms or, in view of (1.122) and (1.127), the operations of lowering and raising indices.

One can also consider different compositions $t \circ \flat$, $t \circ \sharp$ for any tensor t, and thereby can lower and raise indices for arbitrary tensors.

We restrict ourselves to two examples. Take $t \in \mathcal{T}_1^1(M)$ with components $t_{\alpha\cdot}^{\cdot\beta}$, with the places of the indices well indicated. Then we can raise the index α and get a new tensor whose components are, by (1.127),

$$t^{\alpha\beta} = g^{\alpha\lambda}t_{\lambda\cdot}^{\cdot\beta} \tag{1.128}$$

Or we can lower β and get, by (1.122),

$$t_{\alpha\beta} = g_{\beta\lambda}t_{\alpha\cdot}^{\cdot\lambda} \tag{1.129}$$

REMARK In using these operations, the places of the indices must be pointed out clearly.

As a second example, take the tensor g itself. We can raise one of its indices and thereby get a tensor $I \in \mathcal{T}_1^1(M, x)$ given by

$$I(\underline{v}, \omega) = g(\underline{v}, \sharp\omega) \tag{1.130}$$

Because of (1.119), the components of I are δ_α^β, which proves the important fact that the latter defines a tensor: the Kronecker tensor.

Note also that, because of the symmetry of g, it does not matter which index of g we are raising, and the same is true for every symmetric tensor. In the same manner, we can raise both indices of g and get a tensor $\tilde{g} \in \mathcal{T}_0^2(M;x)$ defined by

$$\tilde{g}(\omega_1, \omega_2) = g(\sharp\omega_1, \sharp\omega_2) \tag{1.131}$$

Again using (1.119), we see that the components of $\tilde{g}$ are just $g^{\alpha\beta}$. Hence the $g^{\alpha\beta}$ actually define a tensor of M at x. This tensor is symmetric since $(g^{\alpha\beta})$ is a symmetric matrix.

As a matter of fact, it is worth noting that symmetry properties can be defined in general. Namely, t of (1.108) is symmetric with respect to some arguments if its value remains unchanged by a permutation of those arguments (or indices), whereas it is antisymmetric if its value is unchanged by even permutations, and changes sign under odd permutations.

To get another basic tensor at $x \in M$, let us choose an arbitrary normal vector field $\underline{N}$ defined in a neighborhood of $x \in M$. Then, by (1.75), the action $\underline{v}(\underline{N})$ is defined for every $\underline{v} \in T_xM$. Define

$$b_{\underline{N}}(\underline{v}, \underline{w}) = -\underline{v}(\underline{N}) \cdot \underline{w} \tag{1.132}$$

for every $\underline{v}$, $\underline{w} \in T_xM$, where the dot denotes usual scalar product in $\mathbb{R}^n$. b is clearly a bilinear function on T_xM, and therefore defines a tensor $b \in \mathcal{T}_2^0(M;x)$, which will be called the second fundamental tensor of the couple $(M, \underline{N})$ at x.

The components of the tensor $b_{\underline{N}}$ are

$$b^{\underline{N}}_{\alpha\beta} = b_{\underline{N}}(\underline{x}_\alpha, \underline{x}_\beta) = -\underline{N}_\alpha \cdot \underline{x}_\beta = \underline{N} \cdot \underline{x}_{\beta\alpha} = -\underline{N}_\beta \cdot \underline{x}_\alpha \tag{1.133}$$

where the last equalities follow from

$$\underline{N} \cdot \underline{x}_\alpha = \underline{N} \cdot \underline{x}_\beta = 0$$

by differentiating with respect to u_β and u_α, respectively. Hence $b_{\underline{N}}$ is a symmetric tensor.

If $m = n - 1$ (i.e., in the case of a hypersurface), by orienting a neighborhood of x, we get a well-defined unit normal field $\underline{N}$. The second fundamental tensor defined by this field $\underline{N}$ is denoted simply by b and called the second fundamental tensor of the hypersurface at x.

Let us note that, since both g and b are associated with symmetric bilinear forms, they are also called, respectively, the first and second fundamental forms.

We come back now to general tensors and, inspired by the vector fields, we state

DEFINITION 1.23 A function that associates with every point $x \in M$ a tensor $t \in \mathcal{T}_q^p(M;x)$ is called a tensor field of the type (p, q) on M. The tensor field t is differentiable if the components of its restrictions to ranges of parameterizations are differentiable functions.

Note that the differentiability condition is consistent with the one given for vector fields, and also that it is consistent with the transformation law (1.110), since the changes of parameters are differentiable. Note also that one can define similarly (differentiable) tensor fields on open subsets of M.

Examples: If we consider at every point of M one of the previously defined tensors g, I, b_N, we obviously get differentiable tensor fields. In the case of b_N, this requires using a normal vector field on M, and the tensor field is to be considered only if such a vector field exists. For example, a global second fundamental tensor field for a hypersurface corresponding to a unit field $\underline{N}$ can be obtained for orientable hypersurfaces only.

REMARK The metric tensor field g suggests a very important generalization. Namely, a pair (M, g) consisting of an abstract differentiable manifold M and a symmetric positive definite differentiable tensor field of the type (0, 2) is called a Riemann manifold. Such a tensor field g is called a Riemann metric on M. The Riemann manifold is the basic geometric structure of Einstein's general relativity theory.

Finally, note that as with the notion of the tangent space itself, the theory of tensors has a euclidean character, and this theory can be also used for immersed manifolds, since it depends only on the atlas of the manifold.

EXERCISES

1.51 Write in detailed form the expression $t^{\alpha}_{\beta\alpha}v^{\beta\lambda}$, where the indices are 1 and 2 and where we assume the Einstein convention.

1.52 Prove, using the transformation laws of tensor components, that if $t^{\alpha}_{\beta}v_{\alpha}w^{\beta}$ is a scalar for every vector $\underline{v}$ and every covector w, then t^{α}_{β} are the components of a well-defined tensor of type (1, 1).

1.53 Prove that every linear transformation $\ell : T_xM \to T_xM$ of the tangent space to the manifold M at x yields a well-defined tensor of type (1, 1), and conversely.

1.54 Let A be a tensor of type (1, 1) at $x \in M$, and assume that A has the same components with respect to all the parameterizations of M at x. Prove that these components must be of the form $\alpha\delta^{\mu}_{\lambda}$, $\alpha = \text{const}$. [Hint: Use a change of parameters of the form $\tilde{u}^1 = u^1, \ldots, \tilde{u}^{\lambda} = au^{\lambda}, \ldots, \tilde{u}^m = u^m$ $(a \neq 0, 1)$ to prove that $A^{\mu}_{\lambda} = 0$ for $\lambda \neq \mu$. Next, use a change of the form $u'^1 = u^1, \ldots, u'^{\lambda} = u^{\lambda} + u^{\mu}, \ldots, u'^{\mu} = u^{\lambda} - u^{\mu}, \ldots, u'^m = u^m$, to prove that $A^{\lambda}_{\lambda} = A^{\mu}_{\mu}$.]

1.55 Let t be a tensor of type (0, q) and set

$$t_s(\underline{v}_1, \ldots, \underline{v}_q) = \frac{1}{q!} \sum_\sigma t(\underline{v}_{\sigma(1)}, \ldots, \underline{v}_{\sigma(q)})$$

$$t_a(\underline{v}_1, \ldots, \underline{v}_q) = \frac{1}{q!} \sum_\sigma (\text{sign } \sigma)\, t(\underline{v}_{\sigma(1)}, \ldots, \underline{v}_{\sigma(q)})$$

where σ runs through the set of all the permutations of the indices 1, ..., q. Prove that t_s is a tensor of type (0, q) which is symmetric with respect to all of its arguments, and that t_a is a tensor of type (0, q) which is antisymmetric with respect to all of its arguments. Compute the components of t_s and t_a by means of the components of t.

1.56 Prove that if we lower the upper index of the Kronecker tensor I, we get the metric tensor g.

1.57 Compute the components in an explicit parameterization of the second fundamental tensor of the sphere $S^2 \subset \mathbb{R}^3$ with respect to the unit normal field pointing toward the center of the sphere.

1.58 Let $f: M \to N$ be a differentiable mapping of manifolds, and let t be a differentiable tensor field of type (0, q) on N. Prove that the formula

$$(f_* t)_x(\underline{v}_1, \ldots, \underline{v}_q) = t_{f(x)}(f'_x(\underline{v}_1), \ldots, f'_x(\underline{v}_q))$$

defines a corresponding differentiable tensor field $f_*(t)$ of type (0, q) on M.

1.59 Let t be a differentiable tensor field of type (0, q) on the manifold M, and let $\underline{u}$ be a differentiable tangent vector field. Prove that the formulas

$$(i(\underline{u})t)(\underline{v}_1, \ldots, \underline{v}_{q-1}) = t(\underline{u}, \underline{v}_1, \ldots, \underline{v}_q)$$

$$(L_{\underline{u}} t)(\underline{v}_1, \ldots, \underline{v}_q) = \underline{u}(t(\underline{v}_1, \ldots, \underline{v}_q)) - \sum_{i=1}^{q} t(\underline{v}_1, \ldots, [\underline{u}, \underline{v}_i], \ldots, \underline{v}_q)$$

define two new differentiable tensor fields on M.

1.60 Let ω be a differentiable field of covectors on M, and define

$$(d\omega)(\underline{v}_1, \underline{v}_2) = \underline{v}_1(\omega(\underline{v}_2)) - \underline{v}_2(\omega(\underline{v}_1)) - \omega([\underline{v}_1, \underline{v}_2])$$

Prove that this formula defines a differentiable antisymmetric tensor field of type (0, 2) on M.

2

Curves in E^2 and E^3

2.1 THE NATURAL PARAMETERIZATION

Let C be a differentiable curve embedded in E^3, and $p \in C$. Then if E^3 is referred to a positively oriented cartesian frame, C admits a parameterization

$$\underline{x} = \underline{x}(t) \qquad p = \underline{x}(t_0) \tag{2.1}$$

at p, where (2.1) is a (vector-valued) differentiable function defined for $t \in (a, b)$ which has no singular points, that is,

$$\underline{x}'(t) \neq 0 \qquad t \in (a, b) \tag{2.2}$$

and maps (a, b) homeomorphically onto some open neighborhood of p on C. In the sequel we shall refer to the image of such a parameterization as an <u>arc</u> of C which contains p, and we see that such an arc is the image of a path in C.

We shall be willing to concentrate on arbitrarily small arcs of C containing p, and to study euclidean invariants and invariant properties of such arcs. Such invariants and properties are called <u>local</u>. Obviously, these invariants and properties are characterized by their invariance under change of the cartesian frame in E^3 and under change of the parameterization of C at p.

The first invariance property mentioned above can be checked easily if we use vectorial expressions, since vectors have a geometric character. Concerning the second invariance property, let us recall from Sec. 1.4 that to change the parameterization means to go over from the path (2.1) to a path

$$\underline{x} = \underline{x}(\tilde{t}) \tag{2.3}$$

Here $\tilde{t}$ is a new parameter about p, and on some possibly smaller arc, $\tilde{t}$ is related to t by the transition function

$$\tilde{t} = \tilde{t}(t) \tag{2.4}$$

where

$$\frac{d\tilde{t}}{dt} \neq 0 \tag{2.5}$$

By the inverse function theorem, (2.5) assumes that (2.4) actually is a change of the parameter. That is, we shall have the problem of checking invariance under transformations (2.4) that satisfy (2.5)

First, it is clear that any arc Γ of C is orientable since a continuous field of orientation functions is given by

$$\epsilon(\underline{x}'(t)) = 1 \quad \text{or} \quad \epsilon(\underline{x}'(t)) = -1 \tag{2.6}$$

If we are orienting our arc by choosing one of the two functions of (2.6), and if we study the invariants of this oriented arc, then the relation

$$\frac{d\underline{\underline{x}}}{dt} = \frac{d\underline{\underline{x}}}{d\tilde{t}} \frac{d\tilde{t}}{dt} \tag{2.7}$$

shows that only transformations (2.4) such that

$$\frac{d\tilde{t}}{dt} > 0 \tag{2.8}$$

are allowed.

For example, if we consider the points of Γ to be ordered by the order of the corresponding values of t, then (2.8) shows that this order is an invariant of the oriented arc Γ. If we reverse the orientation, the order is reversed as well. This is also expressed by saying that we chose a sense on Γ.

Therefore, we can divide the invariance check into two parts: First, we shall look for the invariants of the oriented arcs Γ, and then we shall inquire about the behavior of these invariants by reversing the orientation. The second part is easy since it suffices to consider a transformation $\tilde{t} = -t$.

The first part is solved by using a special parameterization.

PROPOSITION 2.1 Let Γ be an oriented arc of C defined by (2.1). Then there are parameterizations $\underline{x} = \underline{x}(s)$ of the arc Γ such that

$$\left| \frac{d\underline{\underline{x}}}{ds} \right| = 1 \tag{2.9}$$

For any two such parameterizations, the corresponding parameters s, $\tilde{s}$ are related by

$$\tilde{s} = s + c \tag{2.10}$$

where c is a constant real number.

Proof: Consider the transformation

$$s(t) = \int_{t_0}^{t} \left| \frac{d\underline{x}}{dt} \right| dt \tag{2.11}$$

This is a differentiable function, and we have

$$\frac{ds}{dt} = \left| \frac{d\underline{x}}{dt} \right| > 0 \tag{2.12}$$

whence we deduce that s is a well-defined and allowed parameter on Γ. Therefore, Γ has a parameterization

$$\underline{x} = \underline{x}(s) \tag{2.13}$$

which is such that

$$\left| \frac{d\underline{x}}{ds} \right| = \left| \frac{d\underline{x}}{dt} \right| \frac{dt}{ds} = 1$$

If for another parameter $\tilde{s}$ we have

$$\left| \frac{d\underline{x}}{d\tilde{s}} \right| = 1 \qquad \frac{d\tilde{s}}{ds} > 0$$

we deduce that

$$1 = \left| \frac{d\underline{x}}{d\tilde{s}} \right| \frac{d\tilde{s}}{ds} = \frac{d\tilde{s}}{ds}$$

which implies (2.10). Q.E.D.

DEFINITION 2.1 A parameter s that satisfies (2.9) is called a canonical or natural parameter.

Proposition 2.1 shows that natural parameters exist for any arc Γ, and they are defined up to a translation. Moreover, if we reverse the orientation of Γ, we get a canonical parameter of the new orientation by setting

$$\bar{s} = -s \tag{2.14}$$

Now, since transformations (2.10) and (2.14) are very simple, it will be easy to check invariance by changing the parameterization when we are using natural parameters. For example, (2.10) yields

$$d\tilde{s} = ds \tag{2.15}$$

whence all the derivatives with respect to s are invariants for the oriented arc Γ.

On the other hand, (2.10) and (2.14) yield

$$d\bar{s}^2 = ds^2 \tag{2.16}$$

for every natural parameter of the nonoriented Γ.

Finally, it is worth recalling the geometric significance of formula (2.11). It is known from elementary calculus that $|s(t)|$, where $s(t)$ is given by (2.11), is the length of the arc of Γ between the points $\underline{x}(t_0)$ and $\underline{x}(t)$. Therefore, $s(t)$ is a curvilinear abscissa on the oriented arc Γ with the origin at $\underline{x}(t_0) = p$. That is, it can be introduced just like the usual abscissa on an oriented straight line.

EXERCISES

2.1 Find natural parameterizations of a straight line, of a circle, of an ellipsis, of a hyperbola, and of a parabola.

2.2 Compute the length of an arc of the curve

$$x^1 = r \cos t \quad x^2 = r \sin t \quad x^3 = ht$$

($t \in \mathbb{R}$, and r and h are constant real numbers) between two fixed points. Write a natural parameterization of this curve. Change the orientation of the curve and write a natural parameterization of the curve with the new orientation

2.3 Repeat Exercise 2.2 for the curve

$$x^1 = e^t \cos t \quad x^2 = e^t \sin t \quad x^3 = e^t$$

2.4 Prove by a straightforward computation that if s is a natural parameter of the curves in Exercises 2.2 and 2.3, the vector $d^2\underline{x}/ds^2$ is an invariant vector of these curves.

2.5 Find a natural parameter for an arc $\underline{x} = \underline{x}(t)$ which is such that $|d\underline{x}/dt| =$ const.

2.2 LOCAL EUCLIDEAN INVARIANTS

A general method for deriving local invariants of manifolds consists of associating an invariant frame with every point of the manifold and expressing the differentials of the components of the frame by the frame itself. This is called the moving frame method, and it works very successfully for arcs Γ of curves in E^3.

Therefore, let Γ be a naturally parameterized oriented arc of an embedded curve C, represented by the equation

$$\underline{x} = \underline{x}(s) \tag{2.17}$$

with $p = \underline{x}(s_0)$.

It is natural to begin the construction of an invariant frame at p by taking as its first vector the positively oriented unit vector of the tangent line of C at p. Since s is a natural parameter, the vector mentioned is defined by

$$\underline{a}(s_0) = \left.\frac{d\underline{x}}{ds}\right|_0 \tag{2.18}$$

Next, define the mapping $\varphi : \Gamma \to S^2$, where S^2 is the unit sphere, by taking $\varphi(s)$ to be the point whose radius vector starting from the center of S^2 is $\underline{a}(s)$ (see Fig. 2.1). We obtain thereby a differentiable path γ of S^2 which, in general, may have singular points. γ is called the tangent spherical image of Γ.

DEFINITION 2.2 A point $p \in \Gamma$ is called an inflection point if the corresponding value s_0 is singular for the paths γ.

Hence an inflection point is characterized by

$$\frac{d\underline{a}}{ds} = 0 \tag{2.19}$$

and we see that this notion is invariant under change of the natural parameter and under change of the frame in $\mathbb{R}^3$.

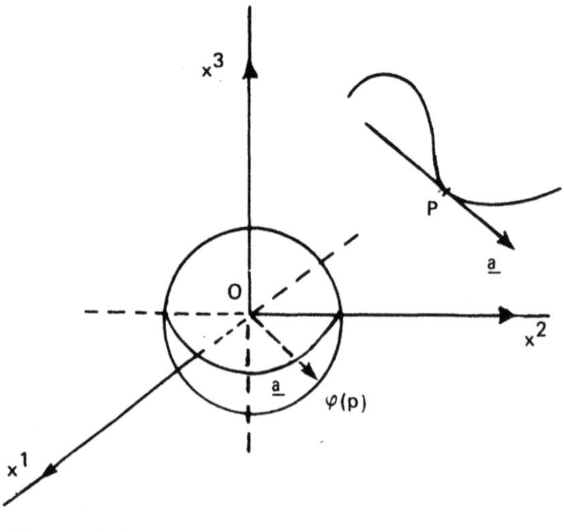

Figure 2.1

PROPOSITION 2.2 If every point of Γ is an inflection point, Γ is a segment of a straight line, and conversely.

Proof: It follows from (2.18) and (2.19) that the hypothesis implies

$$\frac{d\underline{x}}{ds} = \underline{m} = \text{const}$$

which yields

$$\underline{x} = \underline{m}s + \underline{n} \tag{2.20}$$

where $\underline{n}$ is another constant vector. Since (2.20) is the equation of a straight line, the "if" part of Proposition 2.2 is proven. The converse is trivial.

Condition (2.19) also implies that every $p \in \Gamma$ that is not an inflection point has an open neighborhood on Γ which contains no inflection point.

Thereafter we shall assume that p is not an inflection point. Then the velocity vector of γ at s_0 is a nonvanishing vector $\underline{v}(s_0)$ and we have

$$\underline{v}(s_0) = \left.\frac{d\underline{a}}{ds}\right|_0 = \left.\frac{d^2\underline{x}}{ds^2}\right|_0 \tag{2.21}$$

Since $|\underline{a}| = 1$, we get by differentiating $\underline{a}^2 = 1$:

$$\underline{v} \cdot \underline{a} = 0$$

whence $\underline{v} \in N(\Gamma)$, the normal space of Γ at p.

DEFINITION 2.3 The straight line passing through $p \in \Gamma$ whose direction is that of the vector $\underline{v}(s_0)$ is called the principal normal of the curve at p_0. The plane passing through p and containing the tangent line and the principal normal is called the osculating plane of the curve at p. The straight line that passes through p is called the binormal line; obviously, it lies in the normal plane at p. The plane defined by the tangent and the binormal line is called the rectifying plane.

It is obvious that all these are invariant elements and they lead easily to the invariant frame we seek.

Indeed, we can now introduce the unit vectors of the principal normal and of the binormal with a convenient orientation. For example, we can take, respectively,

$$\underline{b} = \frac{\underline{v}}{|\underline{v}|} \qquad \underline{c} = \underline{a} \times \underline{b} \tag{2.22}$$

DEFINITION 2.4 The frame with origin p and with the vector basis $(\underline{a}, \underline{b}, \underline{c})$ is called the Frenet frame of the curve at p.

Furthermore, in order to proceed with the announced method, we have to express the derivatives of $\underline{a}$, $\underline{b}$, $\underline{c}$. We begin with

LEMMA 2.1 Let $\underline{e}_i(s)$ $(i = 1,2,3)$ be three differentiable functions, where $s \in (a,b)$ is some parameter, and where the $\underline{e}_i(s)$ are the vectors of an orthonormal basis in space. Then there is a well-defined skew-symmetric matrix (α_i^j) $(\alpha_i^j = -\alpha_j^i)$ such that

$$\frac{d\underline{e}_i}{ds} = \alpha_i^j \underline{e}_j \tag{2.23}$$

Proof: Since $\{\underline{e}_i\}$ is a vector basis, a unique matrix α_i^j satisfying (2.23) must exist. Namely, this will be the matrix defined by the coordinates of $d\underline{e}_i/ds$ with respect to the basis $\underline{e}_i$. Hence we have to show only that this is a skew-symmetric matrix.

Since $\underline{e}_i$ is orthonormal, we have

$$\underline{e}_i \cdot \underline{e}_j = \delta_i^j \tag{2.24}$$

By differentiating this equation and using (2.23) and (2.24), we get

$$\alpha_i^j + \alpha_j^i = 0$$

Q.E.D.

Now we shall prove

THEOREM 2.1 The Frenet frames of a regular arc Γ which has no inflection points satisfy the following system of linear differential equations:

$$\frac{d\underline{a}}{ds} = \kappa(s)\underline{b} \qquad \frac{d\underline{b}}{ds} = -\kappa(s)\underline{a} + \tau(s)\underline{c} \qquad \frac{d\underline{c}}{ds} = -\tau(s)\underline{b} \tag{2.25}$$

where κ and τ are some differentiable functions of s and $\kappa(s) > 0$.

Proof: The existence of $\kappa(s) > 0$ satisfying (2.25) is a straightforward consequence of formulas (2.21) and (2.22 [note that $\kappa(s) > 0$ is a consequence of our orientation convention for the Frenet frame]. The rest of the theorem follows by Lemma 2.1 and the use of a convenient notation for the coefficients α_i^j of (2.23).

DEFINITION 2.5 Formulas (2.25) are called the Frenet formulas, κ is called the curvature, and τ is called the torsion of the curve at the point $\underline{x}(s)$.

From the very definition it follows that $\kappa(s)$ and $\tau(s)$ are invariants of the oriented arc Γ. As a matter of fact, they are invariants of Γ itself, since, if the orientation is reversed, the signs of s and $\underline{a}$ are changed; therefore, $\underline{b}$ is invariant, $\underline{c}$ changes its sign, and by the last of formulas (2.25), $\tau(s)$ preserves its value and sign.

It is important to note that the Frenet frame and formulas can be introduced in exactly the same way for the more general situation of an immersed arc Γ defined by an equation $\underline{x} = \underline{x}(t)$, which is a differentiable map without singular points, but it is not necessarily a parameterization. Such arcs can appear on immersed curves.

Now, we can prove the following important converse of Theorem 2.1:

THEOREM 2.2 (Fundamental Theorem for Curves in E^3) Let $\kappa(s) > 0$ and $\tau(s)$ be differentiable functions defined on an interval (a,b). Then there are differentiable immersed arcs $\underline{x} = \underline{x}(s)$ defined on (a,b) for which s is a natural parameter, $\kappa(s)$ is the curvature, and $\tau(s)$ the torsion of the arc. Moreover, if Γ_1 and Γ_2 are two such arcs, there is some motion M in E^3 such that $\Gamma_2 = M(\Gamma_1)$.

Proof: Let us start with the Frenet equations (2.25), where $\kappa(s)$ and $\tau(s)$ are the given functions and $\underline{a}(s)$, $\underline{b}(s)$, and $\underline{c}(s)$ are the unknowns. It is a classical result that these equations have unique solutions $\underline{a}(s)$, $\underline{b}(s)$, $\underline{c}(s)$, defined on (a,b) and satisfying the initial conditions

$$\underline{a}(s_0) = \underline{a}_0 \qquad \underline{b}(s_0) = \underline{b}_0 \qquad \underline{c}(s_0) = \underline{c}_0 \tag{2.26}$$

for arbitrarily given vectors $\underline{a}_0$, $\underline{b}_0$, and $\underline{c}_0$.

For these solutions to be candidates for the Frenet frames of the required arc, we must have

$$\begin{aligned} &\varphi_1(s) = \underline{a}^2(s) = 1 \quad \varphi_2(s) = \underline{b}^2(s) = 1 \quad \varphi_3(s) = \underline{c}^2(s) = 1 \\ &\varphi_4(s) = \underline{a}(s)\cdot\underline{b}(s) = 0 \quad \varphi_5(s) = \underline{b}(s)\cdot\underline{c}(s) = 0 \\ &\varphi_6(s) = \underline{c}(s)\cdot\underline{a}(s) = 0 \quad \varphi_7(s) = (\underline{a}(s), \underline{b}(s), \underline{c}(s)) = 1 \end{aligned} \tag{2.27}$$

where $(\underline{a},\underline{b},\underline{c})$ denotes the mixed product $\underline{a}\cdot(\underline{b}\times\underline{c})$, and the last condition expresses the fact that the Frenet frame is positively oriented. We may assume that $\varphi_i(s_0)$ (i = 1, ..., 7) take the values of (2.27); this is obtained by taking $\underline{a}_0$, $\underline{b}_0$, and $\underline{c}_0$ in (2.26) to form an orthonormal and positively oriented basis.

Next, by differentiating the $\varphi_i(s)$ and using (2.25), we get a new system of differential equations:

$$\begin{aligned} &\frac{d\varphi_1}{ds} = 2\kappa\varphi_4 \quad \frac{d\varphi_2}{ds} = -2\kappa\varphi_4 - 2\tau\varphi_5 \quad \frac{d\varphi_3}{ds} = 2\tau\varphi_5 \\ &\frac{d\varphi_4}{ds} = \kappa\varphi_2 - \kappa\varphi_1 - \tau\varphi_6 \quad \frac{d\varphi_5}{ds} = -\kappa\varphi_6 - \tau\varphi_3 + \tau\varphi_2 \\ &\frac{d\varphi_6}{ds} = \kappa\varphi_5 + \tau\varphi_4 \quad \frac{d\varphi_7}{ds} = 0 \end{aligned} \tag{2.28}$$

and it is easy to see that the two sequences of values of (2.27) are solutions of (2.28) and of the corresponding initial conditions at s_0. Hence because such solutions of (2.28) are unique, it follows that the solutions of (2.25) and (2.26) [with $(\underline{a}_0, \underline{b}_0, \underline{c}_0)$ orthonormal] satisfy (2.27).

Now take the equation

$$\frac{dx}{ds} = \underline{a}(s) \tag{2.29}$$

where $\underline{a}(s)$ is the previous solution to (2.25). Integration of (2.29) yields

$$\underline{x}(s) = \underline{x}(s_0) + \int_{s_0}^{s} \underline{a}(s)\, ds \tag{2.30}$$

where $x(s_0) = x_0$ is some prescribed initial value.

Equation (2.30) defines an immersed arc Γ of E^3. By (2.29) the unit tangent vector of Γ is the vector $\underline{a}(s)$, which together with $\underline{b}(s)$ and $\underline{c}(s)$ satisfy (2.25) and (2.27). Hence s is a natural parameter of Γ, $(\underline{a}(s), \underline{b}(s), \underline{c}(s))$ is its family of Frenet frames, and $\kappa(s)$ and $\tau(s)$ are the curvature and the torsion of Γ, respectively. Note that the condition $\kappa(s) > 0$ is necessary, since otherwise we cannot conclude anything about torsion. Thereby the existence part of Theorem 2.2 is proven.

Next, if Γ_1 and Γ_2 are two such arcs, they must be two integrals of (2.25) and (2.29) corresponding to two systems of initial data:

$$I = (\underline{x}_0^1, \underline{a}_0^1, \underline{b}_0^1, \underline{c}_0^1) \qquad II = (\underline{x}_0^2, \underline{a}_0^2, \underline{b}_0^2, \underline{c}_0^2) \tag{2.31}$$

Let M be the motion in E^3 that satisfies $II = M(I)$. Then $M(\Gamma_1)$ and Γ_2 are solutions of (2.25) and (2.29) with the same initial data II, and we must have $\Gamma_2 = M(\Gamma_1)$. Q.E.D.

Since a motion has the same coordinate expression as a change of the cartesian frame in E^3, it follows from Theorem 2.2 that $\kappa(s)$ and $\tau(s)$ actually determine all the local invariants and invariant properties of an immersed curve. In other words, they potentially contain the whole local geometry of the curve, and constitute a complete system of local invariants. In view of this fact, the equations

$$\kappa = \kappa(s) \qquad \tau = \tau(s) \tag{2.32}$$

are called the intrinsic equations of the curve.

REMARK In order to deal with the differential equations (2.25), it suffices to assume $\tau(s)$ continuous and $\kappa(s)$ of class C^1. Then we obtain arcs Γ which are at least of class C^3.

EXERCISES

2.6 Take the natural parameterization of the curve

$$x^1 = e^t \cos t \quad x^2 = e^t \sin t \quad x^3 = e^t$$

obtained in Exercise 2.3 and compute for it the vectors $\underline{a}$, $\underline{b}$, $\underline{c}$, and the invariants κ, τ.

2.7 Consider the curve

$$x^1 = \frac{s}{2}\cos\left(\ln\frac{s}{2}\right) \quad y = \frac{s}{2}\sin\left(\ln\frac{s}{2}\right) \quad z = \frac{s}{\sqrt{2}}$$

Prove that s is a natural parameter, and compute $\underline{a}$, $\underline{b}$, $\underline{c}$, κ, and τ.

2.8 Let Γ be an arc without inflection points, and denote by γ its tangent spherical image. Find a natural parameterization and compute the curvature of γ.

2.9 Prove that the Frenet formulas can be put in the form

$$\frac{d\underline{a}}{ds} = \underline{d}\times\underline{a} \quad \frac{d\underline{b}}{ds} = \underline{d}\times\underline{b} \quad \frac{d\underline{c}}{ds} = \underline{d}\times\underline{c}$$

with a conveniently chosen vector $\underline{d}$. ($\underline{d}$ is called the Darboux vector.)

2.10 Show that the binormal vector $\underline{c}(s)$ of an arc Γ with nowhere vanishing torsion determines $\kappa(s)$ and $|\tau(s)|$.

2.11 Show that the principal normal $\underline{b}(x)$ of an arc Γ with nowhere vanishing torsion determines $\kappa(s)$ and $\tau(s)$.

2.12 Prove that if the arc Γ lies in a plane, the torsion τ vanishes identically on Γ.

2.13 Prove that the normal planes of an arc Γ pass through a fixed point $x_0 \in E_3$ iff Γ lies on a sphere of center x_0.

2.14 Prove that if Γ is an arc such that $\kappa \neq 0$, $\tau \neq 0$, then Γ lies on a sphere iff $\tau/\kappa = (\dot{\kappa}/\tau\kappa^2)^{\cdot}$, where the dot denotes derivative with respect to s. [Hint: Look for the coordinates of the center of the sphere with respect to the Frenet frame of Γ. Use the constancy of the radius and of the center of the sphere.]

2.15 Prove that every arc of constant nonvanishing curvature which lies on a plane or on a sphere is an arc of a circle. [Hint: Use the fundamental theorem for a plane arc. Use Exercise 2.14 to reduce the case of an arc on a sphere to that of a plane arc.]

2.16 Prove that the arcs which have the natural equations $\kappa = \tau = 1/(\sqrt{2}\ s)$ are all obtainable from the arcs of Exercise 2.7 by a motion in space.

2.17 Find the arcs of E^3 that have the natural equations $\kappa = -\tau = 1/(\sqrt{2}\, s)$, and are such that at $t = 1$ one has

$$\underline{a}(1)\left(\frac{1}{\sqrt{2}}, 0, \frac{1}{\sqrt{2}}\right) \quad \underline{b}(1)(0, -1, 0) \quad \underline{c}(1)\left(\frac{1}{\sqrt{2}}, 0, -\frac{1}{\sqrt{2}}\right)$$

2.18 Let $\underline{a}(\alpha_1, \alpha_2, \alpha_3)$, $\underline{b}(\beta_1, \beta_2, \beta_3)$, and $\underline{c}(\gamma_1, \gamma_2, \gamma_3)$ be the Frenet frame of an arc Γ, and set

$$p_i = \frac{\alpha_i + \sqrt{-1}\,\beta_i}{1 - \gamma_i} = \frac{1 + \gamma_i}{\alpha_i - \sqrt{-1}\,\beta_i}$$

$$\bar{p}_i = -\frac{1}{q_i} = \frac{\alpha_i - \sqrt{-1}\,\beta_i}{1 - \gamma_i} = \frac{1 + \gamma_i}{\alpha_i + \sqrt{-1}\,\beta_i} \qquad i = 1, 2, 3$$

Prove that the functions p_i and q_i ($i = 1, 2, 3$) are solutions of the following Riccati equation with complex coefficients

$$\dot{y} = \frac{\sqrt{-1}}{2}\tau y^2 - \sqrt{-1}\,\kappa y - \frac{\sqrt{-1}}{2}\tau$$

Deduce from this that if the general solution of the Riccati equation above is known, the curves of natural equations $\kappa = \kappa(s)$, $\tau = \tau(s)$ can be obtained by quadratures.

2.19 Use the method of Exercise 2.18 to integrate the natural equations of plane curves $\kappa = \kappa(s)$, $\tau = 0$.

2.20 Employ the method of Exercise 2.18 to determine the arcs such that $\kappa = \text{const}$, $\tau = \text{const}$.

2.3 COMPUTATION FORMULAS: COMMENTS AND APPLICATIONS

For concrete applications, computation formulas involving an arbitrary parameterization

$$\underline{x} = \underline{x}(t) \tag{2.33}$$

of an arc Γ are necessary, and such formulas are the first objective of this section.

We recall that the relation between t and the natural parameter s is given by

$$\frac{ds}{dt} = \left|\frac{d\underline{x}}{dt}\right| \tag{2.34}$$

Let us make the convention of denoting by "$'$" the derivatives with respect to t and by "$\cdot$" the derivatives with respect to s. The arc Γ is always supposed to be oriented.

Then, we get, by (2.18),

$$\underline{a} = \frac{\underline{x}'}{|\underline{x}'|} \qquad \frac{d\underline{a}}{ds} = \frac{\underline{x}''}{|\underline{x}'|^2} - \underline{a}\frac{(|\underline{x}'|)'}{|\underline{x}'|^2} \tag{2.35}$$

Here we want to apply the first Frenet formula. Let us note that this formula can be considered true even in the case of an inflection point if we agree to define $\kappa(p) = 0$ if and only if p is an inflection point.

Now the second formula (2.35) gives

$$\kappa\underline{b} = \frac{\underline{x}''}{|\underline{x}'|^2} - \underline{a}\frac{(|\underline{x}'|)'}{|\underline{x}'|^2} \tag{2.36}$$

A first consequence of (2.36) is that $p \in \Gamma$ is an inflection point iff the vector $\underline{x}''$ at p is tangent to Γ at p. That is, $\underline{x}''$ and $\underline{x}'$ are collinear, a condition that can also be expressed by

$$\underline{x}' \times \underline{x}'' = \underline{0} \tag{2.37}$$

If this is not the case, $\underline{b}$ and $\underline{c}$ are well defined, and by left vectorial multiplication of (2.36) by $\underline{a}$ we get

$$\kappa\underline{c} = \frac{\underline{x}' \times \underline{x}''}{|\underline{x}'|^3} \tag{2.38}$$

From this, since $\kappa > 0$, $\underline{c}$ must have the same sense as $\underline{x}' \times \underline{x}''$, and since $\underline{c}$ is a unit vector we obtain

$$\underline{c} = \frac{\underline{x}' \times \underline{x}''}{|\underline{x}' \times \underline{x}''|} \tag{2.39}$$

For the same reasons, (2.38) also yields

$$\kappa = \frac{|\underline{x}' \times \underline{x}''|}{|\underline{x}'|^3} \tag{2.40}$$

a formula that extends to inflection points because of (2.37). It follows that

$$\underline{b} = \underline{c} \times \underline{a} = \frac{(\underline{x}' \times \underline{x}'') \times \underline{x}'}{|(\underline{x}' \times \underline{x}'') \times \underline{x}'|} \tag{2.41}$$

Next, calculating $d\underline{c}/ds$ by using t as an intermediate variable, and using (2.39) and the Frenet formulas, we get

$$-\tau\underline{b} = \frac{1}{|\underline{x}'|}\left[\frac{\underline{x}' \times \underline{x}'''}{|\underline{x}' \times \underline{x}'''|} - \underline{c}\frac{(|\underline{x}' \times \underline{x}''|)'}{|\underline{x}' \times \underline{x}''|}\right] \tag{2.42}$$

whence, by left vector multiplication by $\underline{c}$,

$$\tau\underline{a} = \frac{(\underline{x}' \times \underline{x}'') \times (\underline{x}' \times \underline{x}''')}{|\underline{x}'|\ |\underline{x}' \times \underline{x}''|^2} \tag{2.43}$$

This formula can be simplified by using the well-known identity

$$\underline{u} \times (\underline{v} \times \underline{w}) = (\underline{u} \cdot \underline{w})\underline{v} - (\underline{u} \cdot \underline{v})\underline{w}$$

(whose proof is left to the reader). Indeed, taking $\underline{u} = \underline{x}' \times \underline{x}''$, $\underline{v} = \underline{x}'$, $\underline{w} = \underline{x}'''$, and since $\underline{x}'$ and $\underline{x}' \times \underline{x}''$ are orthogonal, we obtain from (2.43):

$$\tau \underline{a} = \frac{(\underline{x}', \underline{x}'', \underline{x}''')}{|\underline{x}' \times \underline{x}''|^2} \underline{a} \tag{2.44}$$

that is,

$$\tau = \frac{(\underline{x}', \underline{x}'', \underline{x}''')}{|\underline{x}' \times \underline{x}''|^2} \tag{2.45}$$

where, as usually, the parentheses denote the mixed product of vectors.

Finally, let us write the following equations, which are immediate consequences of the formulas obtained:

(a) The tangent line of the curve is defined by

$$\underline{X} = \underline{x} + \lambda \underline{x}' \tag{2.46}$$

where $\underline{X}$ is the radius vector of an arbitrary point of this line.

(b) The osculating plane is defined by $\underline{x}'$ and $\underline{x}''$, in view of (2.36), and it has the equation

$$(\underline{X} - \underline{x}, \underline{x}', \underline{x}'') = 0 \tag{2.47}$$

where $\underline{X}$ is the radius vector of an arbitrary point of the plane.

(c) The equation of the principal normal is

$$\underline{X} = \underline{x} + \lambda(\underline{x}' \times \underline{x}'') \times \underline{x}' \tag{2.48}$$

where λ is a parameter and $\underline{X}$ is an arbitrary point of this normal.

(d) The equation of the binormal is

$$\underline{X} = \underline{x} + \lambda \underline{x}' \times \underline{x}'' \tag{2.49}$$

and so on.

Our next objective is to make some new comments, and to provide some applications of the theory developed.

To begin with, we sketch another interesting interpretation of the osculating plane. Let p be a fixed noninflectional point of the arc Γ defined by (2.33), $p = \underline{x}(t_0)$. Consider the plane π determined by the tangent line of Γ at p and by some other point $\underline{x}(t) \in \Gamma$. This plane contains p, $\underline{x}'(t_0)$, and the vector $\underline{x}(t) - \underline{x}(t_0)$, which can be represented as

$$\underline{x}(t) - \underline{x}(t_0) = (t - t_0)\underline{x}'(t_0) + \frac{(t - t_0)^2}{2} [\underline{x}''(t_0) + \underline{\epsilon}] \tag{2.50}$$

where $\underline{\epsilon} \to 0$ if $t \to t_0$. Therefore, π is defined by p, $\underline{x}'(t_0)$, and $\underline{x}''(t_0) + \underline{\epsilon}$, and one can say that $\{p, \underline{x}'(t_0), \underline{x}''(t_0)\}$ define $\lim_{t \to t_0} \pi$. That is, the osculating plane is $\lim_{t \to t_0} \pi$.

Obviously, this suggests how to define an osculating plane at an inflection point. Moreover, let us consider more general arcs Γ, defined as differentiable paths which are allowed also to have singular points. However, the existing singular points are assumed to satisfy the following conditions:

(a) They are isolated; that is, every such point has a neighborhood with no other singular points.
(b) If (2.33) is the path defining Γ, and t_0 defines a singular point $p = \underline{x}(t_0)$, the following integers are well defined:

(i) The <u>order</u> p, given by

$$\underline{x}'(t_0) = \cdots = \underline{x}^{(p-1)}(t_0) = \underline{0} \quad \underline{x}^{(p)}(t_0) \neq \underline{0} \tag{2.51}$$

(ii) The <u>class</u> q, given by

$$\underline{x}^{(p+1)}(t_0) = \lambda_1 \underline{x}^{(p)}(t_0), \ldots, \underline{x}^{(p+q-1)}(t_0) = \lambda_{p+q-1}\underline{x}^{(p)}(t_0) \tag{2.52}$$

$$\underline{x}^{(p+q)}(t_0) \neq \lambda \underline{x}^{(p)}(t_0)$$

for any λ

(iii) The <u>rank</u> r given by the condition that $\underline{x}^{(p+q+r)}(t_0)$ is the first derivative that is linearly independent of $\underline{x}^{(p)}(t_0)$ and $\underline{x}^{(p+q)}(t_0)$

Clearly, we have $p = 1$ at nonsingular points, $q > 1$ at an inflection point, and so on. At such a singular point, it is natural to define $\underline{a}$, $\underline{b}$, $\underline{c}$, κ, τ by replacing, in formulas (2.35), (2.39), (2.40), (2.41), and (2.45), $\underline{x}'$ by $\underline{x}^{(p)}$, $\underline{x}''$ by $\underline{x}^{(p+q)}$, and $\underline{x}'''$ by $\underline{x}^{(p+q+r)}$.

The following comments concern κ and τ. It is usual in differential geometry to look for geometrical interpretations of the invariants. For example, we have by the first Frenet formula for an oriented arc

$$\kappa = \frac{|d\underline{a}|}{ds} = \frac{d\sigma}{ds} = \lim_{\Delta s \to 0} \frac{\Delta\sigma}{\Delta s} \tag{2.53}$$

where σ denotes the natural parameter on the spherical image γ of Γ. In (2.53) Δs is the variation of the arc length on Γ and $\Delta\sigma$ is the variation of the arc length on γ, and Fig. 2.2 obviously yields

$$\kappa = \frac{d\sigma}{ds} = \frac{d\sigma}{|d\underline{a}|}\frac{|d\underline{a}|}{ds} = \lim_{\Delta s \to 0} \frac{|\tilde{\underline{a}} - \underline{a}|}{\Delta s} = \lim_{\Delta s \to 0} \frac{2 \sin(\Delta\theta/2)}{\Delta s} = \lim_{\Delta s \to 0} \frac{\Delta\theta}{\Delta s} \tag{2.54}$$

where $\Delta\theta$ denotes the angle between the tangent vectors $\tilde{\underline{a}}$ and $\underline{a}$. We leave it to the reader as an exercise to give a similar interpretation for $|\tau|$, beginning with the last Frenet formula. Formula (2.54) will be very important later for plane curves.

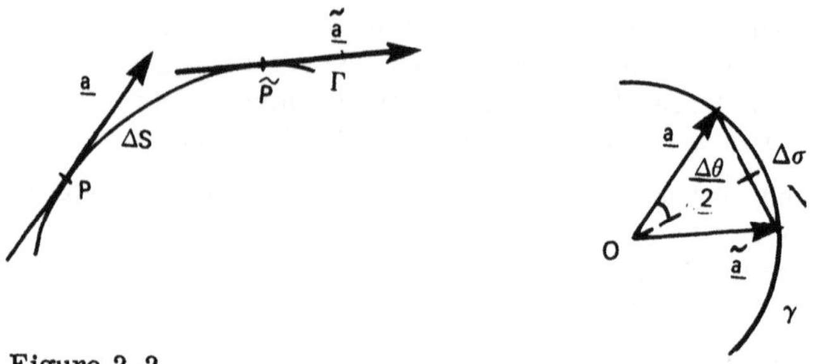

Figure 2.2

There are many other geometrical constructions related to the invariants of a curve. For example:

1. The point

$$\underline{q} = \underline{x}(s) + \frac{1}{\kappa(s)}\,\underline{b} \tag{2.55}$$

is called the center of curvature of the curve at $\underline{x}(s)$. In the case of a plane curve, its locus is called the evolute of the given curve. Conversely, the curve is called the involute of the considered locus; that is, the involutes of a curve are the orthogonal trajectories of the tangents of the curve. (This definition of involutes holds for space curves as well, but in this case the locus of the center of curvature is no longer an evolute, if evolutes are to be taken as curves for which the given curve is an involute.)

2. $\rho(s) = 1/\kappa(s)$ is called the radius of curvature, and $T(s) = 1/\tau(s)$ $[\tau(s) \neq 0]$ is called the radius of torsion.

3. The circle of center $\underline{q}$ and radius $\rho(s)$ is called the circle of curvature (or osculating circle) of Γ at p. And so on.

Now let us consider applications. As a first application of the Frenet formulas, let us provide some information about the graph of the arc $\underline{x} = \underline{x}(s)$ (where s is a natural parameter) in the neighborhood of the regular, noninflectional point $p = x(s_0)$. To do this, let us take as a frame in E^3 the Frenet frame at p, and for the sake of convenience, set $s_0 = 0$. Then we have the following Taylor development:

$$\underline{x}(s) = s\dot{\underline{x}}(0) + \frac{s^2}{2}\ddot{\underline{x}}(0) + \frac{s^3}{6}\dddot{\underline{x}}(0) + \underline{\epsilon} \tag{2.56}$$

where $\underline{\epsilon} \to 0$ together with s^4. In view of the Frenet formulas, we get

$$\dot{\underline{x}}(0) = \underline{a} \quad \ddot{\underline{x}}(0) = \kappa\underline{b} \quad \dddot{\underline{x}}(0) = \kappa^2\underline{a} + \dot{\kappa}\underline{b} + \kappa\tau\underline{c} \tag{2.57}$$

at the point p. Hence we obtain for the coordinates of the point $\underline{x}(s)$ the formulas

$$
\begin{aligned}
x^1 &= s - \kappa^2 \frac{s^3}{6} + \epsilon_1 \\
x^2 &= \kappa \frac{s^2}{2} + \dot{\kappa} \frac{s^3}{6} + \epsilon_2 \qquad (2.58)\\
x^3 &= \kappa\tau \frac{s^3}{6} + \epsilon_3
\end{aligned}
$$

where $\epsilon_i \to 0$ together with s^4 $(i = 1, 2, 3)$.

It follows that for $-\delta < s < \delta$, where $\delta > 0$ and is small enough, the signs of x^1, x^2, x^3 are those of the first terms of (2.58). The projections of the arc on the various planes and the form of the arc are like those shown in Fig. 2.3. These figures also illustrate the difference between the two possible types of curves: those with $\tau > 0$ and those with $\tau < 0$. We can also see that the orientation convention which led to $\kappa > 0$ is equivalent to the fact that the positive sense of the principal normal and the curve are on the same side of the osculating plane. This side is called the concavity of the curve.

A similar method can be employed when discussing the graph of an arc in the neighborhood of a point x_0 of order p, class q, and rank r. To simplify the discussion, we can use here an affine frame (i.e., not necessarily an orthonormal frame) defined by the vectors $\underline{x}_0^{(p)}$, $\underline{x}_0^{(p+q)}$, $\underline{x}_0^{(p+q+r)}$. Then we have the Taylor development

$$
\begin{aligned}
\underline{x}(t) &= \underline{x}(t_0) + \frac{(t - t_0)^p}{p!} \underline{x}_0^{(p)} + \cdots + \frac{(t - t_0)^{p+q}}{(p + q)!} \underline{x}_0^{(p+q)} + \cdots \\
&+ \frac{(t - t_0)^{p+q+r}}{(p + q + r)!} \underline{x}_0^{(p+q+r)} + \underline{\epsilon} \qquad (2.59)
\end{aligned}
$$

and in view of the definition of p, q, r and of the affine frame chosen, we get for the coordinates

$$
\begin{aligned}
x^1 &= \frac{(t - t_0)^p}{p!} + \cdots + \epsilon_1 \\
x^2 &= \frac{(t - t_0)^{p+q}}{(p + q)!} + \cdots + \epsilon_2 \qquad (2.60)\\
x^3 &= \frac{(t - t_0)^{p+q+r}}{(p + q + r)!} + \cdots + \epsilon_3
\end{aligned}
$$

Then we must discuss the sign by means of the parity of p, q, r, and draw the corresponding figures.

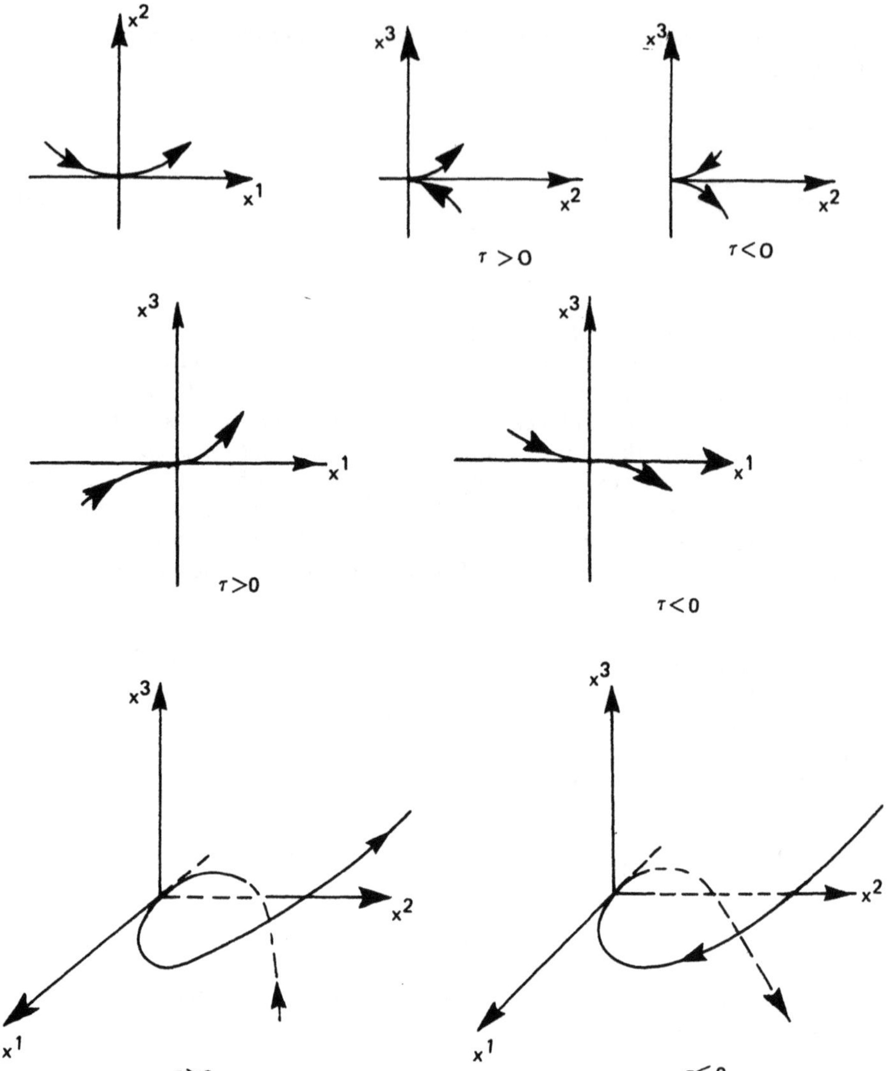

Figure 2.3

As an example, let us consider the possible situations for the projection of the arc on the plane x^1, x^2. These are shown in Fig. 2.4. The four situations are called, respectively:

1. p odd, q odd: geometric regular point
2. p odd, q even: geometric inflection point

3. p even, q odd: cusp of the first kind
4. p even, q even: cusp of the second kind

Note that an inflection point has $p = 1$, $q > 1$, and it is not necessarily a geometric inflection point (unless q is even). However, the name comes from the geometric case.

We see from Fig. 2.4 itself, and because the x^1 axis is the tangent line, that the form of the graph is invariant under changes of the parameterization. The half plane defined by the tangent line and containing the curve in the neighborhood of x_0 is called the concavity of the curve at x_0.

Other applications of the Frenet formulas and the fundamental theorem, which we want to consider are related to intrinsic definitions of various particular curves.

(a) Consider an arc with $\kappa(s) \neq 0$ and $\tau(s) \equiv 0$. Then the last Frenet formula shows that $\underline{c}$ is a constant vector. We have

$$\underline{c}\,\dot{\underline{x}} = \underline{c}\,\underline{a} = 0$$

and by integrating, we get

$$\underline{c}\,\underline{x} = m \tag{2.61}$$

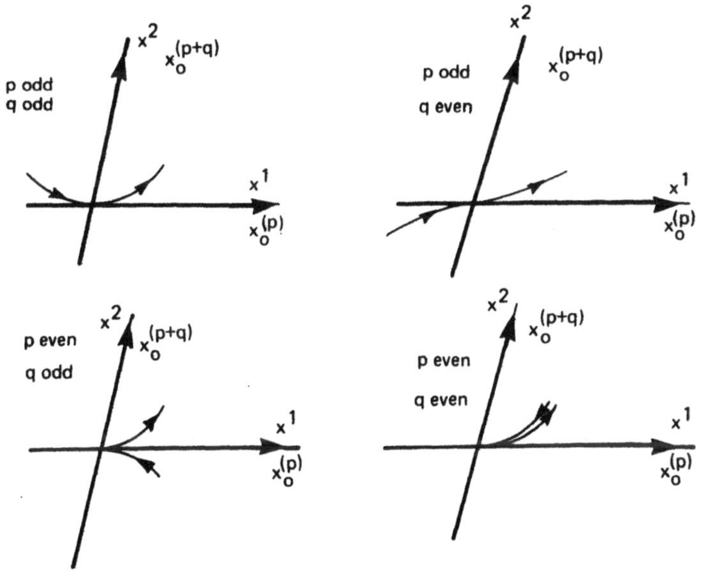

Figure 2.4

where m is a constant number. Since (2.61) is the equation of a plane, it follows that the given arc lies in a plane. The converse is also true since, for a plane arc, $\underline{a}$ and $\underline{b}$ are in the corresponding plane, which is therefore the osculating plane of the curve. It follows that $\underline{c}$ = const, and $\tau = 0$. In view of these results, a point with $\tau = 0$ is called a planar point.

(b) Take the intrinsic equations

$$\kappa = \text{const} \qquad \tau = 0 \tag{2.62}$$

We know that these define plane arcs.

On the other hand, consider the circle of radius $1/\kappa$:

$$x^1 = \frac{1}{\kappa}\cos(\kappa s) \qquad x^2 = \frac{1}{\kappa}\sin(\kappa s) \qquad x^3 = 0 \tag{2.63}$$

where κ is the constant above. It is easy to see that s is a natural parameter and easy to compute the curvature: We find that this curvature is exactly κ. Hence, by the fundamental theorem, every arc that satisfies (2.62) is obtained by moving in space an arc of a circle, and therefore it is itself an arc of a circle.

(c) Take the intrinsic equations

$$\kappa = \text{const} \qquad \tau = \text{const} \neq 0 \tag{2.64}$$

Then we get from the Frenet equations

$$\frac{d\underline{a}}{ds} + \lambda\frac{d\underline{c}}{ds} = \underline{0} \qquad \lambda = \frac{\kappa}{\tau} \tag{2.65}$$

Hence the vector $\underline{a} + \lambda\underline{c}$ is constant, and by drawing through every point of the arc that satisfies (2.64) the line whose direction is that of $\underline{a} + \lambda\underline{c}$, we obtain a cylinder.

If φ denotes the angle between $\underline{a}$ and $\underline{a} + \lambda\underline{c}$, we have

$$\cos\varphi = \frac{1}{1 + \lambda^2} = \text{const}$$

Hence (2.64) defines isogonal trajectories of the generators of a cylinder (i.e., lines cutting the generators at a constant angle). As a matter of fact, this property holds in the more general case $\lambda = \kappa/\tau = \text{const}$.

Conversely, by taking an isogonal trajectory of the generators of a cylinder and letting $\underline{m}$ be a constant vector that defines the direction of these generators, we have $\underline{m}\,\underline{a}$ = const, whence, by differentiation, $\underline{m}\,\underline{b} = 0$, and therefore $\underline{m} = \underline{a} + \lambda\underline{c}$. By differentiating again, and using the Frenet formulas, we get $\kappa/\tau = \lambda$ = const. We have therefore proved that

$$\frac{\kappa}{\tau} = \text{const} \tag{2.66}$$

is the intrinsic characteristic of arcs which are isogonal trajectories of the generators of a cylinder. These are called cylindrical helices.

We now come back to equations (2.64). These clearly define helices, while the first equation suggests trying a circular cylinder in this case. The following arc is thereby suggested:

$$x^1 = r\cos t \quad x^2 = r\sin t \quad x^3 = ht \tag{2.67}$$

where t is not necessarily a natural parameter. This arc lies on the circular cylinder $(x^1)^2 + (x^2)^2 = r^2$. Its tangent direction is $(-r\sin t, r\cos t, h)$, whose angle φ with the z axis has

$$\cos\varphi = \frac{h}{\sqrt{r^2 + h^2}} = \text{const}$$

Using formulas (2.40) and (2.46) we get from (2.67)

$$\kappa = \frac{r}{r^2 + h^2} = \text{const} \qquad \tau = \frac{h}{r^2 + h^2} = \text{const}$$

and we see that r and h can be expressed by κ and τ.

Hence (2.67) is a solution for (2.64) and any other solution will be obtained from (2.67) by a motion in E^3. Since such a motion sends a circular cylinder to a circular cylinder and preserves angles, we see that the arcs defined by (2.64) are just the helices of a circular cylinder. These are called circular helices, and they have many other interesting properties.

We end this section with the remark that the invariants of a curve also have a tensorial interpretation. Namely, if $\underline{x} = \underline{x}(s)$ is an arc with its natural parameterization, its first fundamental tensor, as defined in Sec. 1.8, has as its only component

$$g_{11} = \left(\frac{dx}{ds}\right)^2 = 1$$

with respect to the natural basis $d\underline{x}/ds$. Then if we take the normal field $\underline{b}$, we get a second fundamental tensor (Sec. 1.8) whose only component is

$$b_{11} = -\frac{d\underline{b}}{ds}\cdot\frac{d\underline{x}}{ds} = \kappa$$

Finally, we can define a tensor t by

$$t(\underline{v}, \underline{w}) = (\underline{c}, \underline{v}(\underline{c}), \underline{w})$$

and t has as the unique component

$$c_{11} = \underline{t}\left(\frac{d\underline{x}}{ds}, \frac{d\underline{x}}{ds}\right) = \left(\underline{c}, \frac{d\underline{c}}{ds}, \underline{a}\right) = \tau$$

Therefore, we should expect to get complete systems of invariants of general manifolds as tensor fields. We shall use this method later when studying surfaces.

EXERCISES

2.21 Compute $\underline{a}$, $\underline{b}$, $\underline{c}$, κ, τ for an arc defined by an explicit parameterization $x^2 = x^2(x^1)$, $x^3 = x^3(x^1)$.

2.22 Compute $\underline{a}$, $\underline{b}$, $\underline{c}$, κ, τ for an elementary arc defined by the implicit equations $F(x^1, x^2, x^3) = 0$, $G(x^1, x^2, x^3) = 0$. [Hint: Use equivalent local explicit equations and the results of Exercise 2.21.]

2.23 Compute $\underline{a}$, $\underline{b}$, $\underline{c}$, κ, τ for the following curves:

(a) $x^1 = t$, $x^2 = t^2$, $x^3 = t^3$
(b) $x^1 = \cosh t$, $x^2 = \sinh t$, $x^3 = t$

2.24 Show that the normal planes to the curve

$$x^1 = a \sin^2 t \quad x^2 = a \sin t \cos t \quad z = a \cos t$$

pass through the origin. Compute the curvature and torsion of this curve, and prove, by the criterion of Exercise 2.14, that the curve lies on a sphere. Define this sphere.

2.25 Consider the arc Γ defined by the equations

$$x^1 = \cosh t \quad x^2 = \sinh t \quad x^3 = t$$

and denote by P the intersection point of the tangent line of Γ at $M \in \Gamma$ with the x^1x^2 plane. If M runs through Γ, P describes another arc Γ'. Prove that the tangent line to Γ' at P lies in the osculating plane of Γ at M.

2.26 Assume that the arc Γ has the parameterization

$$x^1 = t \quad x^2 = \sin t \quad x^3 = f(t)$$

Determine the function f(t) such that the principal normals of Γ are parallel to the x^1x^3 plane. Compute the curvature and the torsion of Γ in this case.

2.27 Consider the arc Γ defined by the equations

$$x^1 = t \cos(a \ln t) \quad x^2 = t \sin(a \ln t) \quad x^3 = bt$$

where a, b = const. Prove that the angle between the binormals of Γ and the x^3 axis is constant, and that the principal normals are parallel to the x^1x^2 plane. Compute the curvature and the torsion of Γ.

2.28 Consider the arc Γ defined on the whole real line $\mathbb{R}$ by the equations

$$\underline{x} = \underline{x}(t) = \begin{cases} (t, 0, e^{-1/t^2}) & \text{if } t > 0 \\ (t, e^{-1/t^2}, 0) & \text{if } t < 0 \\ (0, 0, 0) & \text{if } t = 0 \end{cases}$$

(a) Prove that Γ is C^∞, and that $t = 0$ yields an isolated vanishing point of the curvature κ.
(b) Prove that the limit of the osculating plane as $t \to +0$ is $x^2 = 0$, and that the limit of the osculating plane as $t \to -0$ is the plane $x^3 = 0$. (That is, $\underline{b}$ is discontinuous at $t = 0$.)
(c) Show that one can define torsion τ at every point of this arc such that $\tau \equiv 0$ even though Γ is not a plane arc.

2.29 Let Γ be an arc in E^3, and let Γ' be the arc obtained from Γ by the point transformation of E^3 defined by

$$\tilde{x}^i = \sum_{j=1}^{3} a^i_j x^j + b^i \qquad i, j = 1, 2, 3 \qquad (*)$$

where (a^i_j) is an arbitrary nondegenerate 3×3 matrix. Prove that the tangent line and the osculating plane of Γ and Γ' at corresponding points are again corresponding to each other by (*). Can one say the same thing about the other components of the Frenet frames of the two curves?

2.30 Prove that the arc $x^1 = \alpha t$, $x^2 = \beta t^2$, $x^3 = t^3$ belongs to a cylindrical helix iff $4\beta^4 = 9\alpha^2$, and find the direction of the generators of the corresponding cylinder.

2.31 Prove that two arcs Γ and $\tilde{\Gamma}$ differ from each other by a motion and a symmetry with respect to a point iff their curvatures and torsions at symmetric points are related by

$$\tilde{\kappa} = \kappa \qquad \tilde{\tau} = -\tau$$

2.32 Let Γ be an arc of a curve such that $\kappa = \text{const}$, $\tau \neq 0$. Prove that the locus of the centers of curvature of Γ also has a constant curvature.

2.33 Let Γ be an arc that lies on a cone of revolution and cuts the generators at a constant angle. Find equations for the locus Γ' of the centers of curvature of Γ, and prove that Γ and Γ' have orthogonal tangent vectors at corresponding points.

2.34 Find the involutes of a circular helix.

2.35 The arc Γ' is called an evolute of Γ if Γ is an involute of Γ'. If Γ is defined by $\underline{x} = \underline{x}(s)$, where s is a natural parameter, prove that Γ' has an equation $\underline{y} = \underline{x} + \lambda\underline{b} + \mu\underline{c}$, where

$$\lambda = \frac{1}{\kappa} \qquad \mu = \frac{1}{\kappa} \cot\left(\int_a^s \tau \, ds + \text{const}\right)$$

2.36 Prove that an evolute of a plane curve (in space) is a cylindrical helix.

2.37 Prove that the principal normal of a cylindrical helix at a point coincides with the normal to the cylinder at that point.

2.38 Let Γ be an arc of a helix of the cylinder C, and let Γ' be the projection of Γ onto a plane orthogonal to the generators of C. Prove that the curvatures of Γ and Γ' are proportional at corresponding points.

2.39 A curve Γ given by $\underline{x} = \underline{x}(t)$ is called a Bertrand curve if there exists a curve $\tilde{\Gamma}$, $\underline{y} = \underline{y}(t)$, such that Γ and $\tilde{\Gamma}$ have the same principal normal for the same values of t. Then one can put $\underline{y} = \underline{x} + \lambda \underline{b}$. Prove that:
(a) $\lambda = \text{const}$.
(b) Γ is a Bertrand curve iff its curvature and torsion satisfy a linear relation $\alpha\kappa + \beta\tau = 1$ (α, $\beta = \text{const}$).
(c) If Γ has more than one Bertrand mate, it has infinitely many Bertrand mates. This occurs iff Γ is a circular helix.

2.40 Let Γ be an arc represented by $\underline{x} = \underline{x}(s)$, where s is a natural parameter. The sphere that passes through $\underline{x}(s_0)$ and has the center at the point

$$\underline{z} = \underline{x}(s_0) + \rho(s_0)\underline{b}(s_0) + T(s_0)\rho'(s_0)\underline{c}(s_0)$$

($\rho = 1/\kappa$, $T = 1/\tau$) is called the osculating sphere of Γ at s_0. Prove that, if Γ lies on a sphere S, then S is osculating for Γ at every point of Γ.

2.4 GLOBAL THEOREMS FOR EMBEDDED CURVES

Now we proceed with the study of some global properties. It turns out that, in a certain sense, the structure of embedded curves is rather simple. The dimension of the ambient space is irrelevant to this question, and we shall consider an embedded curve C in $\mathbb{R}^n$. Clearly, every point of C has neighborhoods with natural parameterizations, which can be defined as in E^3.

We follow the proofs of J. Milnor (1965).

LEMMA 2.2 Let C be an embedded differentiable curve in $\mathbb{R}^n$, and let Γ_1 and Γ_2 be two arcs of C endowed with the natural parameters s_1 and s_2, respectively. Then $\Gamma_1 \cap \Gamma_2$ has at most two connected components.

Proof: Assume that $\Gamma_1 \cap \Gamma_2 \neq \emptyset$, and let us take the parameterizations $\underline{x} = \underline{x}_i(s_i)$, $s_i \in I_i = (a_i, b_i)$ $(i = 1, 2)$. Then on $\Gamma_1 \cap \Gamma_2$ there is some transition function of the form

$$s_2 = \pm s_1 + c \tag{2.68}$$

To be more precise, because of continuity, a function (2.68) with a well-determined sign and a well-determined constant c is associated with every connected component of $\Gamma_1 \cap \Gamma_2$.

It follows that the graph of the functions (2.68) for the whole of $\Gamma_1 \cap \Gamma_2$ that is, the set

$$G = \{(s_1, s_2) \in I_1 \times I_2 \mid \underline{x}_1(s_1) = \underline{x}_2(s_2)\} \tag{2.69}$$

is a union of segments of slope ± 1 in the s_1, s_2 plane, while there is one such segment for each connected component of $\Gamma_1 \cap \Gamma_2$.

These segments are open in the s_1, s_2 plane, since they are associated with open arcs of $\Gamma_1 \cap \Gamma_2$. But the same segments are closed in the subspace $I_1 \times I_2$ since by (2.69) they are defined by an equality: $\underline{x}_1(s_1) = \underline{x}_2(s_2)$. This is impossible unless the endpoints of the segments are located on the sides of $I_1 \times I_2$. Now, since transition functions are always bijective, it follows that the graph G must look like one of the graphs in Fig. 2.5. Finally, since these graphs contain at most two segments, our lemma is proven.

Actually, Lemma 2.2 means that only the situations in Fig. 2.6 can occur.

The next step is

PROPOSITION 2.3 If the connected embedded curve $C \subset \mathbb{R}^n$ has two naturally parameterized arcs Γ_1 and Γ_2 such that $\Gamma_1 \cap \Gamma_2$ has two connected components, then C is diffeomorphic to the unit circle $S^1 \subset \mathbb{R}^2$.

Proof: If we use the notation of Lemma 2.2 the graph G has two segments which, after a possible change of orientations of the arcs, can be supposed to have slope +1 and to look like that shown in Fig. 2.7. Here we assumed that $a_2 = d$, which clearly can always be obtained by a translation of the natural parameter s_2.

Consider the interval $I = I_1 \cup I_2$ for which we have (see Fig. 2.7)

$$a_1 < c \leq d = a_2 < b_1 = \gamma \leq \delta < b_2 \tag{2.70}$$

Here $b_1 = \gamma$ and

$$c - a_1 = b_2 - \delta \tag{2.71}$$

follow from the fact that the two segments of G have slope 1.

We denote by s the parameter that runs on I; that is, $s = s_1$ on (a_1, b_1) and $s = s_2$ on (a_2, b_2).

Next let us define a map $p: I \to S^1$ by

$$p(s) = \left(\cos\left(\frac{s - a_1}{\delta - a_1} 2\pi\right), \sin\left(\frac{s - a_1}{\delta - a_1} 2\pi\right)\right) \tag{2.72}$$

which is clearly differentiable and onto, since $p((a_1, \delta]) = p([c, b_2)) = S^1$. Finally, define $h: S^1 \to C$ by

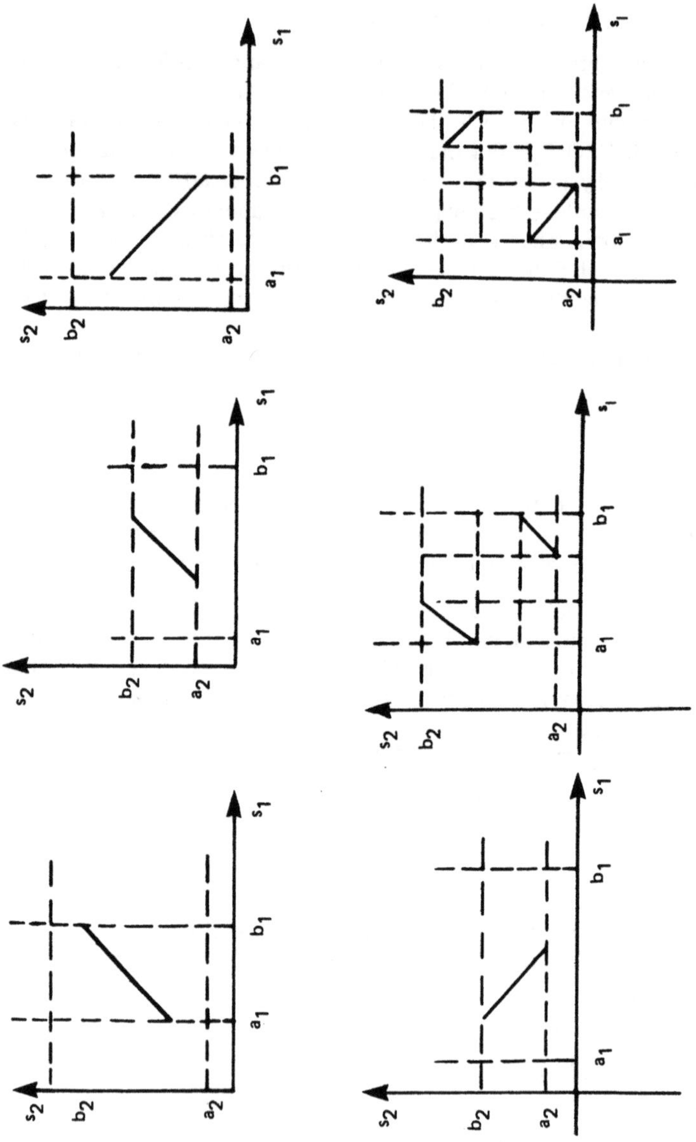

Figure 2.5

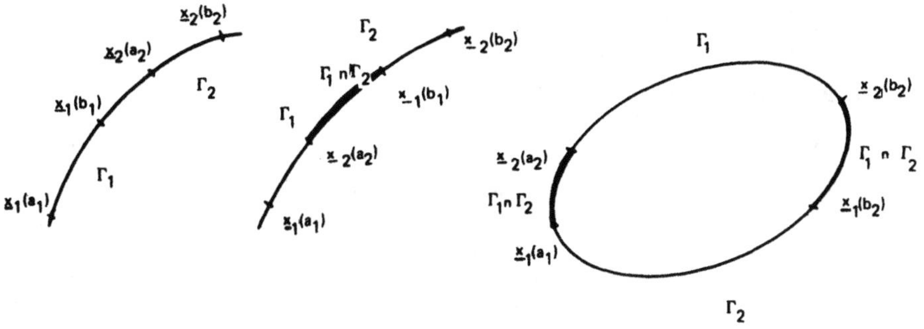

Figure 2.6

$$h(z) = \begin{cases} \underline{x}_1(p^{-1}(z)) & \text{if there is a } p^{-1}(z) \in (a_1, b_1) \\ \underline{x}_2(p^{-1}(z)) & \text{if there is a } p^{-1}(z) \in (a_2, b_2) \end{cases}$$

for every $z \in S^1$.

The function h is well defined. Indeed, for every $z \in S^1$, $p^{-1}(z)$ consists of exactly one value $\sigma_1 \in (a_1, \delta]$ and exactly one value $\sigma_2 \in [c, b_2)$, while

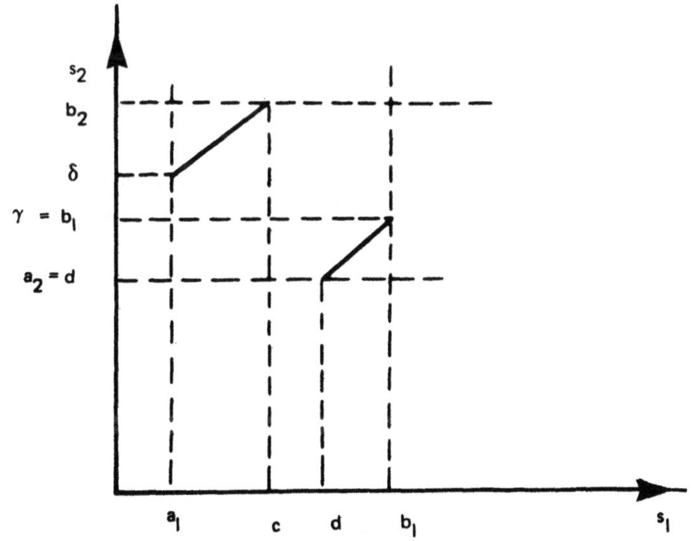

Figure 2.7

$$\frac{\sigma_2 - a_1}{\delta - a_1} 2\pi = \frac{\sigma_1 - a_1}{\delta - a_1} 2\pi + 2k\pi$$

that is, with (2.71),

$$\sigma_2 = \sigma_1 + k(\delta - a_1) = \sigma_1 + k(b_2 - c) \tag{2.74}$$

where k is an integer. Hence either (1) $\sigma_1 = \sigma_2 = \sigma \in [c,\delta]$; or (2) $\sigma_1 \in (a_1,c)$, $\sigma_2 \in (\delta,b_2)$, and $\sigma_2 = \sigma_1 + (\delta - a_1)$. In case (1), either $\sigma \in (a_1,b_1)$, or $\sigma \in (a_2,b_2)$, or σ is a common point of the two intervals, and $\sigma \in (a_2,b_1)$. But in this common part of the two intervals, the transition function (2.68) is exactly $s_2 = s_1$ (Fig. 2.7), whence $\underline{x}_1(\sigma) = \underline{x}_2(\sigma)$ and h(z) is well defined. In case (2), the transition function (2.68) is $s_2 = s_1 + (\delta - a_1)$ (Fig. 2.7) and we again have $\underline{x}_1(\sigma_1) = \underline{x}_2(\sigma_2)$, whence h(z) is well defined.

Moreover, since $\underline{x}_1$ and $\underline{x}_2$ are parameterizations, it also follows that h is an injection. It is also clear that h is a continuous and differentiable mapping between the manifolds S^1 and C.

Since S^1 is compact, it follows that h maps S^1 homeomorphically onto $h(S^1)$, and $h(S^1)$ is a closed subset of C. [Use the topological theorems: (1) a continuous injection of a compact space into a Hausdorff space is a homeomorphism onto the image; and (2) a compact subset of a Hausdorff space is closed.]

On the other hand, $h(S^1) = \Gamma_1 \cup \Gamma_2$ and is therefore open in C. Hence, since C is connected, we must have $h(S^1) = C$, and since h^{-1} is clearly differentiable as well, Proposition 2.3 is proven.

The other possible case is elucidated by

PROPOSITION 2.4 If the embedded curve $C \subset \mathbb{R}^n$ has two naturally parameterized arcs Γ_1 and Γ_2 such that $\Gamma_1 \cap \Gamma_2 \neq \emptyset$, and $\Gamma_1 \cap \Gamma_2$ has one connected component, then $\Gamma_1 \cap \Gamma_2$ is an arc that has a natural parameterization.

Proof: With the notation of Lemma 2.2 the graph G consists of one segment, and the transition function is (2.68) with a well-defined sign and a well-defined constant c. But (2.68) is actually defined on the whole real axis $s_1 \in \mathbb{R}$, whence it can be considered as a transformation of the natural parameter for the whole arc Γ_1. Therefore, $\Gamma_1 \cup \Gamma_2$ is naturally parameterized by s_2, which proves Proposition 2.4.

Now, we can prove the following basic result:

THEOREM 2.3 A connected differentiable curve C embedded in $\mathbb{R}^n$ is diffeomorphic to:

(a) The real line $\mathbb{R}$ if C is noncompact
(b) The unit circle $S^1 \subset \mathbb{R}^2$ if C is compact

Proof: Since C is connected, we can cover it by a countable number of naturally parameterized arcs

$$\Gamma_1, \Gamma_2, \ldots \tag{2.75}$$

such that $(\cup_{h<i} \Gamma_h) \cap \Gamma_i \neq \emptyset \quad (i = 1, 2, \ldots)$.

If $\Gamma_1 \cap \Gamma_2$ has two components, C is diffeomorphic to S^1 by Proposition 2.3. If it has only one component, we can replace $\Gamma_1 \cup \Gamma_2$ by a single arc, by Proposition 2.4, and study the intersection of this arc with Γ_3, and so on.

If C is noncompact, we never arrive at the situation of Proposition 2.3 (since a diffeomorphic image of a circle is compact), and we can conclude that C admits a global natural parameterization. Therefore, C is diffeomorphic to an open segment (a,b) and by a dilatation, it is diffeomorphic to $\mathbb{R}$. [The reader is asked to give a detailed proof of the existence of a diffeomorphism of (a,b) and $\mathbb{R}$.]

On the contrary, if C is compact, the situation above cannot occur, and we must have some intersection of arcs with two components. Then C is diffeomorphic to S^1 by Proposition 2.3. Q.E.D.

Theorem 2.3 yields the classification of the connected embedded curves up to diffeomorphisms. In the case of a nonconnected curve, the components of the curve are determined by the same Theorem 2.3. The compact connected embedded curves are called simple closed curves.

We note also some rather obvious corollaries of Theorem 2.3:

COROLLARY 2.1 Every embedded differentiable curve is orientable.

COROLLARY 2.2 A noncompact embedded curve is always the image of a diffeomorphism $f: (a,b) \to \mathbb{R}^n$. A simple closed differentiable curve is always the image of a differentiable periodic map $f: (a,b) \to \mathbb{R}^n$, where (a,b) is on open interval larger than the period, and the restriction of f to every subinterval shorter than a period is a diffeomorphism. In both cases, f can be taken such as to provide a natural parameterization.

Proof: The first part is obvious. The second part is obvious for S^1 and then we can go over to C by Theorem 2.3.

Therefore, through an abuse of terminology, we can always say that embedded curves have global parameterizations which are either "true" or "periodic." This also suggests considering the following class of immersed curves:

DEFINITION 2.6 A curve C immersed in $\mathbb{R}^n$ is called a geometrically closed curve if it is the image of a differentiable periodic function $f: \mathbb{R} \to \mathbb{R}^n$ without singular points.

Generally, such a curve may have self-intersection points, which does not happen for simple closed curves. In a similar sense, one can consider closed curves with singular points.

We proceed now from another viewpoint with our study of global properties of embedded curves, and come back to three-dimensional euclidean space. From Sec. 2.2 it is obvious that we should expect important results from the study of the invariants of a curve.

Let C be a closed curve in E^3 of total length L. Assuming that we orient C, and consider its curvature function $\kappa = \kappa(s)$, where s is the natural parameter, we can give

DEFINITION 2.7 The total curvature of C is defined by

$$\mu = \int_0^L \kappa(s)\, ds \tag{2.76}$$

By the first Frenet formula, this is just the length of the spherical image γ of the tangents of C. Note that inflection points of C are admitted, and we have $\kappa = 0$ at such a point.

The following interesting result is then available:

THEOREM 2.4 (Fenchel) For every nonsingular closed (not necessarily simple) curve $C \subseteq E^3$, $\mu \geq 2\pi$, and $\mu = 2\pi$ implies that C is a plane curve.

For the following proof, see Chern (1967).

LEMMA 2.3 Let γ be a curve lying on the unit sphere $S^2 \subset \mathbb{R}^3$ which is geometrically closed, differentiable, possibly has singular points, and whose length is well defined and equal to $\ell < 2\pi$. Then there is a point $m \in S^2$ such that the spherical distance $\rho(m, x) \leq \ell/4$ for every $x \in \gamma$. If $\ell = 2\pi$ but γ is not the union of two great semicircles, there is $m \in S^2$ such that $\rho(m, x) < \pi/2$ for all $x \in \gamma$.

Proof: The spherical distance $\rho(a, b)$ is defined as the length of the smallest arc of a great circle that joins a and b. This has the usual properties of a metric: $\rho(a, b) \geq 0$, and $\rho(a, b) = 0$ iff $a = b$; $\rho(a, b) = \rho(b, a)$; $\rho(a, b) + \rho(b, c) \geq \rho(a, c)$. Since S^2 has radius 1, this is actually the radian measure of $\widehat{aOb}$.

If $\rho(a, b) < \pi$, the midpoint m of the arc ab is defined by

$$\rho(a, m) = \rho(b, m) = \frac{1}{2}\rho(a, b)$$

In this case, if $x \in S^2$ and $\rho(m, x) \leq \pi/2$, we can introduce the symmetrical point x' with respect to m on S^2, which implies that

$$2\rho(m, x) \leq \rho(a, x) + \rho(b, x) \tag{2.77}$$

Now let us consider the curve γ and take two points $a, b \in \gamma$ which divide γ into two arcs of equal lengths. It follows that $\rho(a,b) < \pi$ if $\ell < 2\pi$. Let m be the midpoint of a, b, and take a point $x \in \gamma$ such that $2\rho(m,x) < \pi$ (e.g., $x = a$).

Then $\rho(a,x) \leq \ell(a,x)$ and $\rho(b,x) \leq \ell(b,x)$, where ℓ denotes arc length on γ. By (2.77) we get

$$2\rho(m,x) \leq \ell(a,x) + \ell(b,x) = \frac{\ell}{2}$$

Therefore, $f(x) = \rho(m,x)$ $(x \in \gamma)$ is either $\geq \pi/2$ or $\leq \ell/4 < \pi/2$, and since γ is connected and f continuous, only $f(x) \leq \ell/4$ is possible, which proves the first part of the lemma.

Concerning the second part, if $\ell = 2\pi$, then if γ has a pair of antipodal points it must clearly be the union of two great semicircles. If this is not the case, let a, b be as in the case $\ell < 2\pi$, and, moreover, suppose that we may choose them such that

$$\rho(a,x) + \rho(b,x) < \pi \qquad x \in \gamma$$

Then take their midpoint m. If $f(x) = \rho(m,x)$ equals $\pi/2$, for some x_0 we have, by (2.77),

$$2\rho(m,x_0) \leq \rho(a,x_0) + \rho(b,x_0) < \pi$$

that is, $\rho(m,x_0) < \pi/2$, which contradicts $\rho(m,x_0) = \pi/2$. Hence, since $f(a) < \pi/2$ and $\pi/2$ cannot be reached on γ, it follows that $f(x) < \pi/2$ for all $x \in \Gamma$, as stated by the lemma.

Finally, the situation where for the bisecting pairs (a,b) there is a point $x \in \gamma$ such that

$$\rho(a,x) + \rho(b,x) \geq \pi$$

is impossible except in the case of two great semicircles, since it contradicts $\ell = 2\pi$. (The reader is asked to provide the geometric details.)

Lemma 2.3 is thereby completely proven.

Proof of Theorem 2.4: We have to evaluate the length of the spherical image γ of the unit tangent vectors $\underline{a}(s)$ of C. To do this, we shall show that every great circle Γ of S^2 has a nonempty intersection with γ.

Since C is compact, it is bounded, and we can choose a reference frame (O, x^1, x^2, x^3) in E^3 whose plane x^1Ox^2 does not intersect C and is parallel to the plane of Γ.

Let $x^3 = x^3(s)$ be the height function of C with respect to this frame. Because of the compactness of C, $x^3(s)$ must attain a maximum and a minimum value at two points of C, and by elementary calculus, the tangents of C at these points are parallel to x^1Ox^2. Therefore, the corresponding points of γ are on Γ as well.

Now if we had $\mu < 2\pi$ [with μ as in (2.76)], then, by Lemma 2.3, γ would lie in a hemisphere with the pole at the point m of the lemma, and the corresponding equator would not intersect γ. Therefore, $\mu \geq 2\pi$, as stated by Theorem 2.4.

Now let $\mu = 2\pi$. Then, by a similar utilization of Lemma 2.3, γ is seen to be the union of two great semicircles. Hence C is the union of two planar arcs, and since $\underline{a}(s)$ is well defined at every point of C, the latter must actually be a plane curve. Q.E.D.

REMARK However, it is not true that $\mu = 2\pi$ for every closed plane curve. We shall see in the next section which are the closed plane curves such that $\mu = 2\pi$.

COROLLARY 2.3 If C is a closed curve such that $\kappa(s) \leq 1/R$ (R = const), then the length L of C satisfies $L \geq 2\pi R$.

Proof: $L = \int_0^L ds \geq \int_0^L R\kappa(s)\, ds = R \int_0^L \kappa(s)\, ds \geq 2\pi R.$

The importance of the total curvature of a curve lies in its relationship with the position of the curve in the ambient space, in the following sense. We already know that a simple closed curve is diffeomorphic to S^1. Moreover, if it is possible to find in E^3 a continuous family C_t, $t \in [0, 1]$, of simple closed curves such that $C_0 = C$ and $C_1 = S^1$, C is said to be unknotted in E^3. If this is not possible, C is said to be knotted and to define a knot. Figure 2.8 gives an idea of these situations.

We note that the topological theory of knots is one of the important chapters of algebraic topology. Knots are related with total curvature by

THEOREM 2.5 (Fary-Milnor) For every simple closed knotted curve, the total curvature satisfies $\mu \geq 4\pi$.

The proof of this theorem is beyond our scope. However, let us try to explain why the stated result should be true. In fact, we must prove that if $2\pi \leq \mu < 4\pi$, C is unknotted. In the proof of Fenchel's theorem we saw that

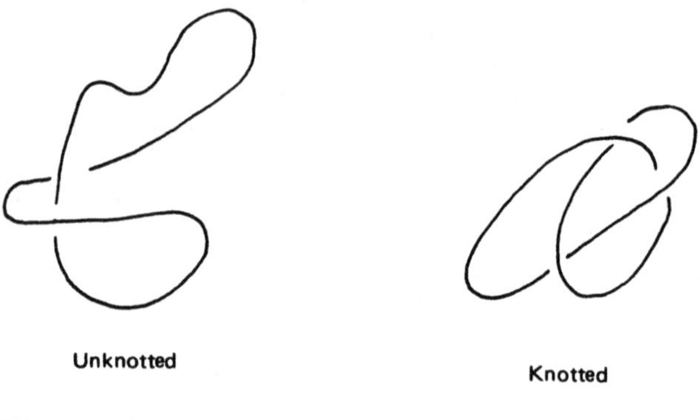

Figure 2.8

every great circle of S^2 crosses the spherical image γ of C at two points, which may or may not be different. One can prove that, for $\mu < 4\pi$, there actually is a great circle Γ on S^2 which has no more intersection points with γ. The height function with respect to the plane of Γ then has one minimum and one maximum point on C, and every plane parallel to the plane of Γ, which is between these two points, cuts C in two points. The corresponding chords of C generate a surface which is homeomorphic to a closed disk and whose boundary in C. The necessary continuous deformation of C can be performed in this disk, and therefore C is unknotted. [See a proof of Theorem 2.5 in Chern (1967).]

EXERCISES

2.41 Give an explicit construction of a diffeomorphism between an open segment (a, b) and the real line $\mathbb{R}$.

2.42 Provide an explicit construction of a diffeomorphism between an ellipse and the unit circle S^1.

2.43 Prove by using ideas involved in the proof of Fenchel's theorem that the tangent spherical image of an ellipse is a great circle of the unit sphere S^2.

2.44 Using Fenchel's theorem, prove that the curve $x^1 = a \sin t + b \cos t$, $x^2 = a \sin t - b \cos t$, $x^3 = -2a \sin t$ is a closed plane curve. What kind of a curve is it? Use Corollary 2.3 to obtain a lower bound for the length of the curve.

2.45 Consider the piecewise differentiable simple closed curve defined by

$$\underline{x}(t) = \begin{cases} (\cos t,\ \sin t,\ t) & t \in [0, 2\pi] \\ (1,\ 0,\ 4\pi - t) & t \in [2\pi, 4\pi] \end{cases}$$

(which can be extended periodically to $t \in \mathbb{R}$). One can prove that the Fary-Milnor theorem holds for such curves as well. Use it in order to prove that the defined curve $\underline{x}(t)$ is unknotted.

2.46 Prove that any real number r can be obtained as $\int_0^L \tau(s)\, ds$, where $\tau(s)$ is the torsion of some closed curve in space, with natural parameter s and total length L. [Hint: Construct a curve consisting of a convenient number of coils of a convenient circular helix, and of a plane curve that joins the endpoints of the helix.]

2.5 PLANE CURVES

The theory of curves in the plane E^2, which was considered in previous sections as a particular case, has interesting independent aspects, some of which will be pointed out here. If

$$\underline{x} = \underline{x}(s) \qquad s \in (a,b) \tag{2.78}$$

is a plane arc, it suffices to attach it a Frenet frame in the plane only, and we could define this frame by means of the unit vectors $\underline{a}$ and $\underline{b}$ of Sec. 2.2, which obviously belong to the plane of the arc.

But then it may happen that the orientation of the vector basis $(\underline{a},\underline{b})$ will be negative with respect to the plane. One can avoid this by replacing $\underline{b}$ with a unit vector $\underline{e}$ such that $\underline{e} = \underline{b}$ if the orientation of $(\underline{a},\underline{b})$ is positive, and $\underline{e} = -\underline{b}$ if the orientation of $(\underline{a},\underline{b})$ is negative. (Recall that the positive orientation in a plane is in general defined by the intuitive convention that it is the counterclockwise sense.)

As a result, we have the following plane Frenet formulas:

$$\frac{d\underline{a}}{ds} = k\underline{e} \qquad \frac{d\underline{e}}{ds} = -k\underline{a} \tag{2.79}$$

where $|k| = \kappa$, and it is now k(s) which will be called the curvature of the plane curve. By comparison with the fundamental theorem in E^3, it follows that k(s) defines the oriented plane arc up to a displacement in the plane. Note that k(s) changes its sign if the orientation of the arc is reversed. k = k(s) is called the natural equation of a plane arc.

It is also worthwhile to remark that if for two plane arcs we have $\kappa_1(s) = \kappa_2(s)$, they can be obtained from each other by a displacement in space, but possibly not in the plane, where they could, for instance, be symmetric with respect to a line.

We shall not discuss the equation of the tangent line or the principal normal line (called here simply the normal line) of a plane curve in detail. These can be obtained by particularizing the known general formulas. (For example, the normal line appears here as the normal line of a hypersurface, since a curve in E^2 is a hypersurface of E^2!) Let us simply note these equations:

$$\underline{X} = \underline{x} + \lambda\underline{x}' \qquad \text{the tangent line} \tag{2.80}$$

$$\underline{x}' \cdot (\underline{X} - \underline{x}) = 0 \qquad \text{the normal line} \tag{2.81}$$

and so on.

But we do want to emphasize the curvature k(s) a bit more. First, let us look for its geometrical meaning.

In the case of a plane curve, we have a map $p : \Gamma \to S^1$ defined by $p(x(s)) = \underline{a}(s)$, which is continuous and differentiable. Let us consider the angle between $\underline{a}(s)$ and the x^1 axis in E^2. There is a uniquely determined value $\alpha(s) \in [0, 2\pi)$ for the measure of this angle, but the function $\alpha(s)$ is not continuous at the points s_0 where $\alpha(s_0) = 0$. This difficulty is overcome by

PROPOSITION 2.5 Let $[a', b'] \subset (a, b)$ be any closed interval. Then there are continuous functions $\theta(s)$ on $[a', b']$ such that, for every $s \in [a', b']$, $\theta(s) = \alpha(s) + 2m(s)\pi$, where $m(s)$ is an integer. Such a function is determined by prescribing $m(a')$ arbitrarily. Two such functions differ by $2h\pi$ for some constant integer h.

Proof: Since $[a', b']$ is closed, $\underline{a}(s)$ is uniformly continuous, and there is a real number $\delta > 0$ such that $|s' - s| < \delta$ implies that $\underline{a}(s')$ belongs to the open semicircle with midpoint $\underline{a}(s)$. Let us choose a partition of $[a', b']$ by points $a' < s_1 < \cdots < s_n < b'$ such that $|s_{i+1} - s_i| < \delta$ $(i = 1, \ldots, n-1)$, and $|s_1 - a'| < \delta$, $|b' - s_n| < \delta$.

Define $\theta(a')$ by taking an arbitrary integer for $m(a')$. Then there is a unique $\theta(s)$ for $s \in [a', s_1]$ which is continuous and is of the form $\alpha(s) + 2k(s)\pi$ [$k(s)$ is an integer]. Indeed, it suffices to take the value of $\theta(s)$ such that

$$|\theta(s) - \theta(a')| < \frac{\pi}{2}$$

which is possible by the choice of the partition. This value is unique since

$$|\theta - \theta(a')| < \frac{\pi}{2} \qquad |\theta' - \theta(a')| < \frac{\pi}{2}$$

implies that $|\theta' - \theta| < \pi$, and the two values θ and θ' cannot measure the same angle. The continuity of this $\theta(s)$ is obvious.

This first step fixed $\theta s_1)$, and with the value obtained we can proceed similarly on $[s_1, s_2]$, and so on. The function $\theta(s)$ is thereby defined in a finite number of steps. If $\theta(s)$ and $\theta'(s)$ are two such functions, $\theta'(s) = \theta(s) + 2h(s)\pi$, where $h(s)$ is continuous and assumes integer values. Hence $h(s)$ is a constant integer, and this ends the proof of Proposition 2.5.

Hence we can fix in a continuous manner the measure of the angle between the oriented tangent line and the x^1 axis. Then the coordinates of the vector $\underline{a}$ are

$$\underline{a}(\cos\theta, \sin\theta) \tag{2.82}$$

Since from $\underline{a} = d\underline{x}/ds$ we also have

$$\underline{a}\left(\frac{dx^1}{ds}, \frac{dx^2}{ds}\right) \tag{2.83}$$

it follows that

$$\frac{dx^1}{ds} = \cos\theta \quad \frac{dx^2}{ds} = \sin\theta \tag{2.84}$$

and, locally,

$$\theta = \arc\tan\frac{\dot{x}^2}{\dot{x}^1} = \arc\tan\frac{(x^2)'}{(x^1)'} \tag{2.85}$$

Hence $\theta(s)$ is differentiable in this case. (As in Sec. 2.3, the dot denotes the derivative with respect to s, and the prime denotes the derivative with respect to an arbitrary parameter t.)

From (2.82) and from the definition of $\underline{e}$ we also get

$$\underline{e}(-\sin\theta, \cos\theta) \tag{2.86}$$

Then differentiating (2.82) and comparing with the first formula (2.79) yields

$$k(s) = \dot{\theta}(s) = \frac{d\theta}{ds} = \lim_{\Delta s \to 0} \frac{\Delta\theta}{\Delta s} \tag{2.87}$$

Formula (2.87) provides us with the geometric interpretation of the curvature, which is similar to that of κ in the case of a space curve, where a function θ was not available. Moreover, here $\Delta\theta$ is oriented, which also fixes the sign of the curvature. Another consequence of the established results is a computation formula for k(s). Indeed, from (2.85) and (2.87), we get

$$k(s) = \frac{(x^1)'(x^2)'' - (x^2)'(x^1)''}{[(x^1)'^2 + (x^2)'^2]^{3/2}} \tag{2.88}$$

Finally, we now have a very easy way to integrate the natural equation $k = k(s)$ of a plane curve. Namely, Eq. (2.87) yields

$$\theta(s) = \theta(s_0) + \int_{s_0}^{s} k(s)\, ds \tag{2.89}$$

where $\theta(s_0)$ is an arbitrary value.

Next, we get by (2.84),

$$\begin{aligned} x^1(s) &= x^1(s_0) + \int_{s_0}^{s} \cos\theta(s)\, ds \\ x^2(s) &= x^2(s_0) + \int_{s_0}^{s} \sin\theta(s)\, ds \end{aligned} \tag{2.90}$$

where $x^1(s_0)$ and $x^2(s_0)$ are arbitrary constants.

At this point we should like to add some further comments about the function $\theta(s)$. First, as a matter of fact, only the local existence of the continuous function $\theta(s)$ has been used up to this stage. Therefore, it would have sufficed to define $\theta(s)$ by (2.85) in a neighborhood of a point, and (2.87) would remain valid at that point.

Now let us start with an arc Γ, $\underline{x} = \underline{x}(s)$, defined on (a, b), and compute its curvature k(s). Then use (2.89) and (2.90) to define a new arc $\tilde{\Gamma}$, $\underline{x} = \tilde{\underline{x}}(s)$, such that $\tilde{x}^1(s_0) = x^1(s_0)$, $\tilde{x}^2(s_0) = x^2(s_0)$, $\theta(s_0)$ = the local $\theta(s_0)$ of Γ. The fundamental theorem yields $\tilde{\Gamma} = \Gamma$, and we obtain thereby a new proof of Proposition 2.5 for the whole of (a, b). Namely, the function required by Proposition 2.5 will be the $\theta(s)$ of (2.88). Another remark is that all the previous results are true for immersed arcs as well.

Proposition 2.5 and the results about k(s) lead to interesting global theorems for closed plane curves, and we shall study a few of these theorems. Let C be a closed immersed curve in E^2, defined as the image of a periodic differentiable function $\underline{x} : \mathbb{R} \to E^2$, whose equation is (2.78) and for which s is the natural parameters and ℓ the total length. Consider a continuous function $\theta(s)$ as in Proposition 2.5, defined on $[0, \ell]$. Then, because of periodicity, there is a well-defined integer n such that

$$\theta(\ell) - \theta(0) = 2n\pi$$

and this integer does not depend on the choice of θ. The number n is called the <u>rotation index</u> of C. Note that in view of the relation between θ and k we get

$$n = \frac{1}{2\pi} \int_C k(s)\, ds \tag{2.91}$$

The following theorem is intuitively clear, and we want to sketch for it a mathematical proof:

THEOREM 2.6 (Theorem of Turning Tangents) The rotation index of a simple closed curve is ± 1.

We give here only a sketchy explanation of the theorem. A full proof can be found in Chern (1967).

By an important topological theorem (the Jordan theorem), any plane simple closed curve C defines a plane region called its <u>interior</u>. Moreover, there is a diffeomorphism sending the curve C and its interior onto a circular disk.

Hence if we make a circular hole in the interior of C, the rest behaves like a ring, and we see that we can include C in a continuous family of curves C_t, $t \in [0, 1]$, such that $C_0 = C$, C_1 is a circle, and every C_t is a differentiable simple closed curve. (A plane simple closed curve is unknotted!)

Now, by (2.91), $n(C_t)$ is expressed by an integral depending continuously on the parameter t. Hence $n(C_t)$ is a continuous function of t which assumes only integer values, and we must have $n(C_t) = \text{const}$. But $n(C_1)$ is clearly ± 1, where the sign depends on the orientation of the circle C_1. It follows that $n(C_0) = n(C) = \pm 1$. Q.E.D.

We shall use Theorem 2.6 to get a characterization of the convex curves.

DEFINITION 2.8 A plane curve C is called convex if C lies entirely in one of the closed half planes defined by the tangent line at x, for every $x \in C$.

A nonsimple curve is clearly nonconvex, whence a convex curve is simple (embedded curve). One can prove that a differentiable plane curve, which is the boundary of a convex plane region is convex in the sense of Definition 2.8, and if C is a plane simple closed convex curve, its interior (which exists by the Jordan theorem) is a convex region (i.e., whenever it contains the points p, q it contains the whole segment [p, q]).

THEOREM 2.7 A differentiable closed curve C in the plane is convex if and only if it is simple and its curvature k does not change sign on C.

Proof: Since $k(s) = \dot{\theta}(s)$, $s \in [0, \ell]$ (where ℓ is the total length of C), it follows that k does not change its sign iff θ is monotonic. Let C be convex. We know that it is simple.

Suppose that $\theta(s_1) = \theta(s_2)$ for $s_1 < s_2$. Then $\underline{a}(s_1) = \underline{a}(s_2)$. On the other hand, Theorem 2.6 implies that $\theta(s)$ takes every value in $[0, 2\pi]$ and there is an s_3 such that $\underline{a}(s_3) = -\underline{a}(s_1)$. But a convex curve cannot have three distinct parallel tangents (since one of them will be between the two others and therefore it separates the curve). Hence we have one line t which is tangent to C at two distinct points p, q. Again using the convexity of C, we see that the whole segment [p, q] belongs to C. Hence the common tangent has a well-defined orientation, and the points p, q must actually be $\underline{x}(s_1)$, $\underline{x}(s_2)$. Moreover, along the line segment $[\underline{x}(s_1), \underline{x}(s_2)]$ $\underline{a}(s) = \text{const}$, hence $\theta(s) = \text{const}$.

Hence if θ is, for instance, nondecreasing in a neighborhood of some initial point s_0, it must be nondecreasing everywhere. Indeed, if at some point $s_1 > s_0$, θ begins to decrease, we get two neighboring points $s_1' < s_1 < s_1''$ with $\theta(s_1') = \theta(s_1'')$. But this would imply $\theta(s) = \text{const}$ on $[s_1', s_1'']$, in contradiction to the decreasing hypothesis.

A similar situation occurs for a nonincreasing θ. Hence we have the general conclusion that θ must be monotonic. The conditions stated are therefore necessary for C to be a convex curve.

Conversely, let C be simple and, by choice of a convenient orientation, $k(s) \geq 0$. Then $\theta(s)$ is monotonically nondecreasing, and runs from 0 to 2π when $s \in [0, \ell]$, in view of Theorem 2.6. It follows that if $\underline{a}(s_1) = \underline{a}(s_2)$ for $0 \leq s_1 < s_2 < \ell$, $\underline{a}(s) = \text{const}$ on $[s_1, s_2]$ and $C \mid [s_1, s_2]$ is a line segment along which we have a fixed tangent line.

Now if C is not convex, let us choose s_0 such that C has points on both sides of the tangent t of C at $x(s_0) = p$. Consider the function $h: C \to \mathbb{R}$

which measures the distance from the points of C to the line t, with the sign + on one side of t and - on the other side. Then h is a continuous function, and since C is compact it has a maximum and a minimum point, which clearly belong to different half planes.

The tangents t_1, t_2 at these points are parallel to t, and at least two of the lines t, t_1, t_2 have the same orientation. But by the arguments above, this is impossible unless C is entirely in one half plane of t. This contradiction shows that C must be convex. Q.E.D.

Let us return for a moment to Fenchel's theorem (Theorem 2.4). If C is a plane closed convex curve, then by Theorem 2.7, k(s) does not change sign and we may assume that $k(s) = \kappa(s)$ (after a convenient orientation). Next, by Theorem 2.6 the total curvature of C is $\mu = \int_C \kappa(s)\, ds = 2\pi$. Conversely, in Theorem 2.4 we proved that if $\mu = 2\pi$, the simple closed curve C is a plane curve. So, by Theorem 2.6 we have

$$\int_C k(s)\, ds = \int_C \kappa(s)\, ds = 2\pi$$

and since $|k(s)| = \kappa(s)$, this happens iff $k(s) = \kappa(s) \geq 0$. Therefore, by Theorem 2.7, C is convex.

That is, Fenchel's theorem can be completed by the assertion that the total curvature of a simple closed curve is $\mu = 2\pi$ iff the curve is a plane convex curve.

Appendix

Let us add to this section some remarks about the graph of a plane curve which are familiar to the reader but are worth recalling, nevertheless. It is often necessary to construct the graph of a plane arc*

$$x = x(t) \qquad y = y(t) \tag{2.92}$$

The functions x, y are assumed to define a differentiable arc, which may have isolated singular points of order p and class q.

In the first place, we shall look for information offered directly by the functions (2.92): domains, intersections with the coordinate axes, interesting limits, and so on. In particular, it is important to find the asymptotes, which are straight lines d such that the distance $\delta(p, d)$ from the point p(t) of the arc to d has the limit 0 for $t \to t_0$ (or $t \to \pm\infty$), where $p(t) \to \infty$ as $t \to t_0$.

The asymptotes should be found exactly as in the case of arcs $y = y(x)$ studied in elementary calculus. That is, if $\lim_{t\to t_0} x(t) = a$ and $\lim_{t\to t_0} y(t) = \infty$, $x = a$ is an asymptote. If $\lim_{t\to t_0} x(t) = \infty$ and $\lim_{t\to t_0} y(t) = a$, $y = a$ is an asymptote. If $\lim_{t\to t_0} x(t) = \infty$, $\lim_{t\to t_0} y(t) = \infty$, and if the following

*For the sake of convenience here we set $x^1 = x$, $x^2 = y$.

limits exist

$$m = \lim_{t \to t_0} \frac{y(t)}{x(t)} \qquad n = \lim_{t \to t_0} [y(t) - mx(t)] \tag{2.93}$$

the line $y = mx + n$ is an asymptote.

Next, one looks for information offered by the first derivatives: the variation and the extreme points of $x(t)$ and $y(t)$; singular points with their order and class, whose parity shows the form of the curve. Then recall that for $y = y(x)$, the position of the curve with respect to its tangent is characterized by the sign of d^2y/dx^2. A simple computation yields

$$\frac{d^2y}{dx^2} = \frac{x'y'' - y'x''}{x'^2} \tag{2.94}$$

which is therefore the quantity that allows us to study the concavity and the inflection points of the curve.

Finally, one puts together all this information for $-\infty < t < +\infty$, and draws the graph.

A similar problem can be considered for

$$F(x, y) = 0 \tag{2.95}$$

where F is differentiable and possibly has isolated singular points. At nonsingular points, there are local solutions [e.g., $y = y(x)$] whose derivatives are

$$y' = -\frac{F_x}{F_y} \qquad y'' = -\frac{1}{F_y^3}\begin{vmatrix} F_{xx} & F_{xy} & F_x \\ F_{xy} & F_{yy} & F_y \\ F_x & F_y & 0 \end{vmatrix} \tag{2.96}$$

and similar information can be derived. The method of studying singular points consists of intersecting (2.95) with a line

$$y - y_0 = \lambda(x - x_0) \tag{2.97}$$

and looking for local parametric equations of (2.95) at (x_0, y_0), where λ is the parameter. Asymptotes can be determined by a transformation

$$x = \frac{1}{\xi} \qquad y = \frac{\eta}{\xi} \tag{2.98}$$

followed by the computation of the limits m and n of (2.93) in the new variables and for $\xi \to 0$.

Often, the following practical device is helpful. One tries to put (2.95) in the form

$$\Phi_1\Phi_2 \cdots \Phi_h = \Psi_1\Psi_2 \cdots \Psi_k \tag{2.99}$$

where $\Phi_i(x, y) = 0$ and $\Psi_j(x, y) = 0$ have easily representable graphs. Then one eliminates the region of the plane where the signs of the two parts of (2.99) are different. The remaining regions give the form of the graph.

EXERCISES

2.47 Let π be a parabola, O its vertex, and d the tangent line of π at O. Find all the arcs C that lie in the plane of π and are such that the midpoint Q of the segment MP, where $M \in C$ and P is the intersection point of d with the normal line of C at M, belongs to π. [Hint: Choose a frame with the origin at O and the x^1 axis d. Then represent C by $y = f(x)$.]

2.48 Find the plane arcs C with the property $OM^2 = d/\kappa$, where O is a fixed point in the plane, $M \in C$, κ is the curvature of C at M, and d is the distance from O to the tangent line of C at M. [Hint: Represent C by $y = f(x)$.]

2.49 Introduce polar coordinates (ρ, φ) in E^2 by defining

$$x^1 = \rho \cos \varphi \qquad x^2 = \rho \sin \varphi$$

When does the equation $\rho = \rho(\varphi)$ represent a regular plane arc? In this case, write the equations of the tangent and of the normal line. Compute the curvature k of the arc considered.

2.50 A circular disk of radius 1 in the x^1x^2 plane rolls without slipping along the x^1 axis. Then the locus described by a fixed point of the circumference of the disk is a curve called a cycloid. Find a parameterization of the cycloid. Compute the length of an arc of the cycloid which corresponds to a complete rotation of the disk. Compute the curvature k of the cycloid.

2.51 Consider the equations

$$x^1 = \cos t \qquad x^2 = \cos t + \ln \tan \frac{t}{2}$$

Prove that they define a differentiable arc with one singular point at $t = \pi/2$. Draw the graph of this curve; it is called the tractrix. Prove that the length of the segment of the tangent of the tractrix between the point of tangency and the x^2 axis is constantly equal to 1. Compute the curvature k at an arbitrary point of the tractrix.

2.52 Consider the curve

$$x^1 = \cosh t \qquad x^2 = t \qquad t \in \mathbb{R}$$

(called the catenary). Draw its graph. Determine the evolute of this curve. Show that the tractrix is an involute of the catenary.

2.53 Find the plane curves of the natural equation:
(a) $k = 1/(\alpha s + \beta)$ $(\alpha, \beta = \text{const})$
(b) $k = \alpha/(\alpha^2 + s^2)$ $(\alpha = \text{const})$

2.54 Find the plane curves that satisfy the equation

$$\rho^2 + \alpha^2 = \alpha^2 \exp\left(-\frac{2s}{\alpha}\right)$$

where ρ is the radius of curvature, s the natural parameter, and $\alpha = \text{const}$.

2.55 The points of a curve where $dk/ds = 0$ are called vertices. Prove that an ellipse has exactly four vertices. (A famous theorem states that a plane closed convex curve has at least four vertices. This is called the four-vertex theorem.)

2.56 Prove that if a closed plane curve C is contained inside a disk of radius r, there is at least one point $p \in C$ such that $\kappa(P) \geq 1/r$. [Hint: Prove that the stated inequality actually holds for the smallest concentric disk that still contains C.]

2.57 Let C be a plane closed curve with the rotation index $n > 0$ and with total length ℓ. Assume that the curvature of C satisfies the inequality $k \leq M$. Prove that $\ell \geq 2\pi n/M$.

2.58 Let C be an oriented plane closed curve with curvature $k > 0$ and with at least one self-intersection point. Prove that the rotation index of C is at least 2.

3

Surfaces in E^3

3.1 THE FUNDAMENTAL TENSORS

The purpose of this chapter is to study the main differential invariants of a surface in Euclidean space E^3. This case contains the basic ingredients for the whole of modern differential geometry, and it is essentially distinct from the case of the curves since no simple canonical invariant parameterization is now available. However, we have seen in Sec. 2.3 that the invariants of a curve have a tensor interpretation, and therefore it will be natural to look for invariant tensors in the case of the surfaces as well.

Generally, the surfaces considered can be either embedded or immersed. As a rule, we consider first embedded surfaces, and define their geometric elements by using a parameterization. Then the same elements will be considered automatically on immersed surfaces by attaching them to every elementary part of the surface.

Let S be an embedded surface, that is, an embedded two-dimensional manifold of E^3, where the latter is referred to a positively oriented cartesian frame with coordinates x^i ($i = 1, 2, 3$). Then there are at every point $p \in S$ parameterizations

$$\underline{x} = \underline{x}(u^\alpha) \qquad \alpha = 1, 2 \tag{3.1}$$

of the form studied in Sec. 1.4, and these will be used, as in Chap. 1, to express the various elements related to the surface S as a differentiable manifold. Note that implicit or explicit equations can also be used. (See Secs. 1.3 and 1.4.) For instance, the vectors $\underline{x}_\alpha$ ($= \partial\underline{x}/\partial u^\alpha$) define a basis of the tangent space, the tangent plane is given by Eqs. (1.48) and (1.49), the normal line is defined as in Sec. 1.5 and so on. In particular, let us emphasize that $\underline{x}_\alpha$ ($\alpha = 1, 2$) are noncollinear vectors [since (3.1) are the equations of a diffeomorphism; see Sec. 1.2] which can be expressed, using the vector product, by

$$\underline{x}_1 \times \underline{x}_2 \neq 0 \tag{3.2}$$

Every parameterization (3.1) has two associated families of paths defined, respectively, by $u^1 = u^1$, $u^2 = \text{const}$, and $u^1 = \text{const}$, $u^2 = u^2$. These are called the parametric lines of (3.1) and together they form the parametric net of the parameterization. The composition of the map (3.1) with these paths defines curves in E^3 which are tangent to the vectors $\underline{x}_1$ and $\underline{x}_2$, respectively.

The following proposition is often useful.

PROPOSITION 3.1 Let $\underline{u}$ and $\underline{v}$ be two tangent vector fields defined on an open subset $W \subset S$, and nowhere collinear in W. Then one can find at every point $p_0 \in W$ a parameterization $\underline{x}(u^1, u^2)$ such that $\underline{u} = a\underline{x}_1$, $\underline{v} = b\underline{x}_2$ (a, b are real nonvanishing functions).

Proof: Let us start with any parameterization $\underline{y} = \underline{y}(t^\alpha)$ ($\alpha = 1, 2$) at p_0 such that p_0 is $\underline{y}(t_0^\alpha)$. We shall find a change of this parameterization: $u^\alpha = u^\alpha(t^\beta)$ such that for the new parameterization $\underline{x} = \underline{x}(u^\alpha)$ the desired conclusion holds. This means that the components of $\underline{u}$ and $\underline{v}$ with respect to the natural basis of the parameterization $\underline{x}$ must be $a\delta_1^\alpha$ and $b\delta_2^\alpha$, (α = 1, 2; δ is the Kronecker index, respectively.

If, on the other hand,* $\underline{u} = \xi^\beta \underline{y}_\beta$, $\underline{v} = \eta^\beta \underline{y}_\beta$, the transformation law (1.100) yields

$$\xi^\beta = a\delta_1^\alpha \frac{\partial t^\beta}{\partial u^\alpha} = a \frac{\partial t^\beta}{\partial u^1}$$

$$\eta^\beta = b\delta_2^\alpha \frac{\partial t^\beta}{\partial u^\alpha} = b \frac{\partial t^\beta}{\partial u^2}$$

This implies that

$$dt^1 = \frac{1}{a} \xi^1 \, du^1 + \frac{1}{b} \eta^1 \, du^2$$

$$dt^2 = \frac{1}{a} \xi^2 \, du^1 + \frac{1}{b} \eta^2 \, du^2$$

and by solving these equations (which is possible since $\xi^1\eta^2 - \xi^2\eta^1 \neq 0$ because $\underline{u}$ and $\underline{v}$ are noncollinear),

$$\begin{aligned} du^1 &= a(\lambda \, dt^1 + \mu \, dt^2) \\ du^2 &= b(\gamma \, dt^1 + \sigma \, dt^2) \end{aligned} \tag{3.3}$$

where λ, μ, γ, and σ are known differentiable functions defined in some open neighborhood of (t_0^α).

---

*The reader is reminded that we always use the Einstein summation convention explained in Sec. 1.8.

Now it is well known in calculus that every form $\alpha\, dt^1 + \beta\, dt^2$ has an integral multiplier f, which is defined in some open neighborhood of (t_0^α) and for which $f(\alpha\, dt^1 + \beta\, dt^2)$ is an exact differential. [Indeed, it suffices, for example, to find the general integral $\Phi(t^1, t^2) = \text{const}$ of the equation $dt^2/dt^1 = -\alpha/\beta$. Then $\alpha\, dt^1 + \beta\, dt^2 = 0$ is equivalent to $d\Phi = 0$, that is, $d\Phi = f(\alpha\, dt^1 + \beta\, dt^2)$.]

Hence a and b can be found such that (3.3) holds, and by integrating (3.3), we obtain the desired change of the parameterization, whose jacobian does not vanish because $\underline{u}$ and $\underline{v}$ are noncollinear. Q.E.D.

Now we proceed to a consideration of the basic tensors of S at p.

As shown in Sec. 1.8, we have a first fundamental tensor (or metric tensor) g defined by the scalar product in E^3, and which is therefore a euclidean invariant of S at p. We shall use the notation and properties of this tensor as developed in Sec. 1.8, while the Greek indices there will now assume the values 1, 2. Note that over the whole of S, g defines a differentiable tensor field.

A classical notation for the components of g is

$$g_{11} = E \quad g_{12} = g_{21} = F \quad g_{22} = G \tag{3.4}$$

It is also usual to call g the first fundamental form of S, and E, F, G are the coefficients of this form. Note that the tensor itself, and not the coefficients E, F, G, is invariant.

As indicated by its name, the metric tensor is used to perform measurements on S. Thus the length of a tangent vector $\underline{v} \in T_pS$ is given by

$$|\underline{v}| = [g(\underline{v}, \underline{v})]^{\frac{1}{2}} = (g_{\alpha\beta} v^\alpha v^\beta)^{\frac{1}{2}} \tag{3.5}$$

and the angle of $\underline{v}, \underline{w} \in T_pS$ is given by

$$\cos(\widehat{\underline{v}, \underline{w}}) = \frac{g(\underline{v}, \underline{w})}{|\underline{v}|\,|\underline{w}|} = \frac{g_{\alpha\beta} v^\alpha w^\beta}{(g_{\alpha\beta} v^\alpha v^\beta)^{\frac{1}{2}} (g_{\lambda\mu} w^\alpha w^\mu)^{\frac{1}{2}}} \tag{3.6}$$

where $\underline{v} = v^\alpha \underline{x}_\alpha$, $\underline{w} = w^\alpha \underline{x}_\alpha$ with respect to the parameterization $\underline{x}$.

Moreover, assume that S is orientable (or at least, that we are restricting ourselves to an orientable neighborhood of $p \in S$), and choose a unit normal vector field $\underline{N}$. (See Sec. 1.7.) Then we can speak of orientable angles in T_pS, the orientation being defined by that of E^3, and by $\underline{N}$ in such a manner that if φ is the oriented angle between $\underline{v}$ and $\underline{w}$, one has

$$\underline{v} \times \underline{w} = |\underline{v}|\ |\underline{w}|\ \sin\varphi \cdot \underline{N} \tag{3.7}$$

In other words,

$$\sin \varphi = \frac{(\underline{N}, \underline{v}, \underline{w})}{|\underline{v}|\ |\underline{w}|} \tag{3.8}$$

Now, if we set $\epsilon(\underline{v}, \underline{w}) = (\underline{N}, \underline{v}, \underline{w})$, we clearly get a two-times covariant tensor (or tensor field, if we are varying p), which is antisymmetric and has the components (see Sec. 1.8)

$$\epsilon_{\alpha\beta} = \epsilon(\underline{x}_\alpha, \underline{x}_\beta) = (\underline{N}, \underline{x}_\alpha, \underline{x}_\beta) \tag{3.9}$$

Since the $\underline{x}_\alpha$ are tangent to S, we can take

$$\underline{N} = \frac{\underline{x}_1 \times \underline{x}_2}{|\underline{x}_1 \times \underline{x}_2|}$$

and by a well-known identity, we have

$$|\underline{x}_1 \times \underline{x}_2| = [\underline{x}_1^2 \underline{x}_2^2 - (\underline{x}_1 \cdot \underline{x}_2)^2]^{\frac{1}{2}} = (\det g)^{\frac{1}{2}} \tag{3.10}$$

where $\det g = \det (g_{\alpha\beta}) = g_{11}g_{22} - g_{12}^2$. Hence

$$\underline{N} = \frac{\underline{x}_1 \times \underline{x}_2}{(\det g)^{\frac{1}{2}}} \tag{3.11}$$

and the components of the tensor ϵ are

$$\epsilon_{11} = \epsilon_{22} = 0 \qquad \epsilon_{12} = -\epsilon_{21} = (\det g)^{\frac{1}{2}} \tag{3.12}$$

Therefore, this tensor depends on the metric tensor only, and we can write instead of (3.8),

$$\sin \varphi = \frac{\epsilon_{\alpha\beta} v^\alpha w^\beta}{(g_{\alpha\beta} v^\alpha v^\beta)^{\frac{1}{2}} (g_{\lambda\mu} w^\lambda w^\mu)^{\frac{1}{2}}} \tag{3.13}$$

the notation being as in (3.6).

Furthermore, let us consider a differentiable path $\gamma : (a,b) \to S$, whose image lies in the range of the parameterization $\underline{x}$. Then $\underline{x} \circ \gamma$ is a curve in E^3 and the length of the arc of this curve which corresponds to $[c,d] \subset (a,b)$ is given by (Sec. 2.1)

$$\ell = \int_c^d \left| \frac{d\underline{\underline{x}}}{dt} \right| dt$$

where t is the parameter on (a,b). But

$$\frac{d\underline{\underline{x}}}{dt} = \underline{x}_\alpha \frac{du^\alpha}{dt} = \underline{x}_\alpha (u^\alpha)' \tag{3.14}$$

is tangent to S, whence we have by (3.5),

$$\ell = \int_c^d [g_{\alpha\beta}(u^\alpha)'(u^\beta)']^{\frac{1}{2}}\, dt \tag{3.15}$$

That is, if we are referring to such curves as curves on the surface S, we see that the length of such curves are computable by means of the first fundamental tensor of the surface.

In particular, we can use (3.15) to introduce the natural parameter s (Sec. 2.1) for curves on S, and we get for it

$$ds^2 = \Phi = g_{\alpha\beta}\, du^\alpha\, du^\beta \tag{3.16}$$

with differentials computed along the given path. Formula (3.16) is a classical version of the first fundamental form of a surface, and ds^2, Φ is classical notation for this form. For paths that do not lie in the range of a parameterization, the length can be computed by dividing them into pieces for which (3.15) is applicable and adding the lengths of these pieces.

Note again that (3.14) gives the coordinates of the tangent vector to a curve on S with respect to the natural bases $\underline{x}_\alpha$. This enables us to define the angle between two curves on S, both passing through the point p, as the angle of their tangent directions, and to compute it with the help of the first fundamental tensor g.

For example, if we denote by α the angle between the parametric lines of the parameterization $\underline{x}$, we get from (3.6) and (3.13)

$$\cos\alpha = \frac{g_{12}}{(g_{11}g_{22})^{\frac{1}{2}}} \qquad \sin\alpha = \frac{(\det g)^{\frac{1}{2}}}{(g_{11}g_{22})^{\frac{1}{2}}} \tag{3.17}$$

Hence $g_{12} = 0$ characterizes the parameterizations with an orthogonal parametric net.

Another important metric notion is that of the area of a domain $D \subset S$. If D belongs to the range of a parameterization $\underline{x}$, we shall define its area by

$$A(D) = \iint_{\mathscr{D}} (\det g)^{\frac{1}{2}}\, du^1\, du^2 \tag{3.18}$$

where $\mathscr{D} = \underline{x}^{-1}(D)$. The reader is asked to prove as an exercise that this formula is invariant under changes of the parameterization. (Use the change-of variables formula in a double integral!) If D is larger, we shall divide it into pieces contained in ranges of parameterizations and then add the areas of these pieces, thereby computing the total area of D.

Of course, there are good reasons for formula (3.18), reasons that are usually explained in books on calculus [see, e.g., Kaplan (1968)]. Hence we shall not insist here on this matter.

Of course, at this point, the reader will not be surprised if we also consider the second fundamental tensor of a surface as defined in Sec. 1.8. But it would be interesting to arrive at this tensor by a geometric process, which will be described here. Once again, suppose that we have an oriented unit normal field $\underline{N}$ given by (3.11). Let us denote by S^2 the unit 2-sphere, with the center at the origin and radius 1. Then we can define a natural mapping $\gamma : S \to S^2$ by sending the point $p \in S$ to the point with radius vector $\underline{N}(p)$ of S^2. γ is called the Gauss map or the spherical image of S.

Let us assume for a moment that $\underline{N}_1 \times \underline{N}_2 \neq \underline{0}$ ($\underline{N}_\alpha = \partial\underline{N}/\partial u^\alpha$!) in the neighborhood of p parameterized by Eq. (3.1). Then $\underline{x} = \underline{N}(u^\alpha)$ ($\alpha = 1, 2$) is a parameterization of S^2 at $\gamma(p)$, and the local analytic representation of γ with respect to these parameterizations (Sec. 1.6) is

$$u^\alpha = u^\alpha \quad \alpha = 1, 2 \tag{3.19}$$

Hence, in this case, γ is differentiable, and it has a differential $\gamma'_p : T_pS \to T_{\gamma(p)}S^2$, which sends the vector $\xi^\alpha \underline{x}_\alpha$ to the vector $\xi^\alpha \underline{N}_\alpha$. But since the tangent plane of the sphere is orthogonal to the radius vector at the same point, the vectors $\underline{N}_\alpha$ can be regarded as vectors of T_pS as well and γ'_p can be considered as a linear transformation of the linear space T_pS. (See Fig. 3.1.)

Moreover, using the transformation laws of Sec. 1.8, we see that the correspondence γ'_p defined by $\xi^\alpha \underline{x}_\alpha \longmapsto \xi^\alpha \underline{N}_\alpha$ makes sense at every point $p \in S$ (even if $\underline{N}_1 \times \underline{N}_2 = \underline{0}$). Indeed, the chain rule for partial derivatives shows that the $\underline{N}_\alpha$ satisfy formula (1.94) while the ξ^α satisfy (1.100); thus $\xi^\alpha \underline{N}_\alpha$ does not depend on the parameterization and is a well-defined tangent vector. Hence we can give

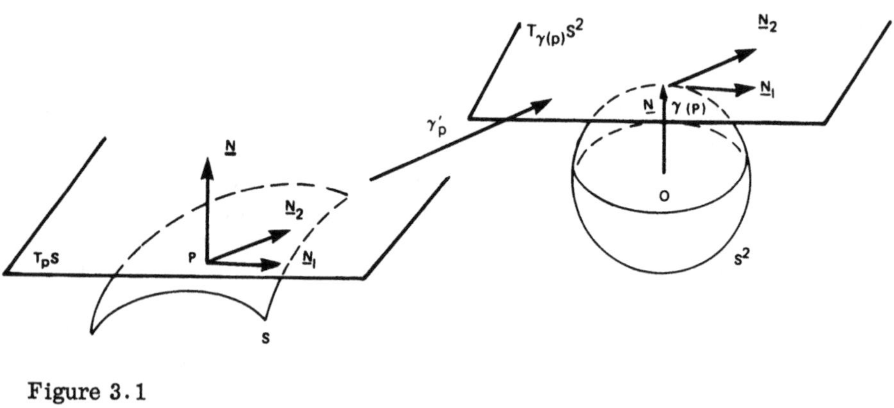

Figure 3.1

DEFINITION 3.1 The linear transformation $\ell = -\gamma'_p : T_pS \to T_pS$ is called the Weingarten transformation.

Since $\underline{N}_\alpha \in T_pS$, there are some coefficients b^β_α such that

$$\underline{N}_\alpha = -b^\beta_\alpha \underline{x}_\beta \tag{3.20}$$

(The signs are a matter of convenience only.) Then if we denote $\ell(\xi^\alpha \underline{x}_\alpha) = \xi^\beta \underline{x}_\beta$, we get

$$\xi^\beta \underline{x}_\beta = -\xi^\alpha \underline{N}_\alpha = b^\beta_\alpha \xi^\alpha \underline{x}_\beta$$

and we see that ℓ has the coordinate expression

$$\xi^\beta = b^\beta_\alpha \xi^\alpha \tag{3.21}$$

with respect to the natural basis $\underline{x}_\alpha$ of T_pS. Since b^β_α are obviously differentiable functions, the Weingarten transformations over S send differentiable vector fields to differentiable vector fields.

The formulas (3.20) are classically known as the Weingarten formulas.

PROPOSITION 3.2 The Weingarten transformation is self-adjoint, that is, it satisfies the relation

$$\ell\underline{v} \cdot \underline{w} = \underline{v} \cdot \ell\underline{w} \tag{3.22}$$

Proof: If $\underline{v} = \xi^\alpha \underline{x}_\alpha$, $\underline{w} = \eta^\beta \underline{x}_\beta$, then $\ell(\underline{v}) = -\xi^\alpha \underline{N}_\alpha$, $\ell(\underline{w}) = -\eta^\beta \underline{N}_\beta$, and we have

$$\ell\underline{v} \cdot \underline{w} = -(\underline{N}_\alpha \cdot \underline{x}_\beta)\xi^\alpha \eta^\beta \qquad \underline{v} \cdot \ell\underline{w} = -(\underline{N}_\beta \cdot \underline{x}_\alpha)\xi^\alpha \eta^\beta$$

But from $\underline{N} \cdot \underline{x}_\beta = 0$ we get by differentiation

$$-\underline{N}_\alpha \cdot \underline{x}_\beta = \underline{N} \cdot \underline{x}_{\alpha\beta} = -\underline{N}_\beta \cdot \underline{x}_\alpha$$

This implies (3.22). Q.E.D.

It follows from Proposition 3.2 that the formula

$$b(\underline{v}, \underline{w}) = \ell\underline{v} \cdot \underline{w} = \underline{v} \cdot \ell\underline{w} \tag{3.23}$$

defines a symmetric bilinear map $b : T_pS \times T_pS \to \mathbb{R}$, hence a symmetric two-times covariant tensor at $p \in S$. The components of this tensor are

$$b_{\alpha\beta} = b(\underline{x}_\alpha, \underline{x}_\beta) = \ell\underline{x}_\alpha \cdot \underline{x}_\beta = -\underline{N}_\alpha \underline{x}_\beta = -\underline{N}_\beta \underline{x}_\alpha = \underline{N} \cdot \underline{x}_{\alpha\beta} \tag{3.24}$$

By comparing this with formula (1.133), we see that we obtained just the second fundamental tensor (or form) of S at p as defined in Sec. 1.8. Consequently, we also have (1.132), that is,

$$b(\underline{v}, \underline{w}) = -\underline{v}(\underline{N}) \cdot \underline{w} = -\underline{w}(\underline{N}) \cdot \underline{v} \tag{3.25}$$

whence by comparing with (3.23),

$$\ell(\underline{v}) = -\underline{v}(\underline{N}) \tag{3.26}$$

which is another interesting expression for the Weingarten transformation. Formula (3.26) is also called the Weingarten formula or the Weingarten equation. If p varies on S, b defines a tensor field which will also be referred to as the second fundamental tensor or form of S.

Classically, the components of the tensor b are denoted by

$$b_{11} = L \quad b_{12} = b_{21} = M \quad b_{22} = N \tag{3.27}$$

From (3.11) and (3.24) we get the following computation formulas:

$$b_{\alpha\beta} = \frac{(\underline{x}\ ,\underline{x}_2,\underline{x}_{\alpha\beta})}{(\det g)^{\frac{1}{2}}} \qquad \alpha,\ \beta = 1,\ 2 \tag{3.28}$$

Furthermore, in order to compute the Weingarten transformation, we consider $\underline{v} = \xi^{\alpha}\underline{x}_{\alpha}$, $\underline{w} = \eta^{\beta}\underline{x}_{\beta}$, and

$$\begin{aligned} b(\underline{v}, \underline{w}) &= b_{\alpha\beta}\xi^{\alpha}\eta^{\beta} = \ell(\underline{v}) \cdot \underline{w} = (b^{\gamma}_{\alpha}\xi^{\alpha}\underline{x}_{\gamma}) \cdot (\eta^{\beta}\underline{x}_{\beta}) \\ &= g_{\beta\gamma}b^{\gamma}_{\alpha}\xi^{\alpha}\eta^{\beta} \end{aligned}$$

whence

$$b_{\alpha\beta} = g_{\beta\gamma}b^{\gamma}_{\alpha} \tag{3.29}$$

Now we can consider the contravariant tensor $g^{\alpha\beta}$ of Sec. 1.8, which is related to the first fundamental tensor of S by (1.119),

$$g_{\alpha\lambda}g^{\lambda\beta} = \delta^{\beta}_{\alpha} \tag{3.30}$$

Here α, β, $\lambda = 1,\ 2$, and by solving these equations we get

$$g^{11} = \frac{g_{22}}{\det g} \quad g^{12} = -g^{21} = \frac{g_{12}}{\det g} \quad g^{22} = \frac{g_{11}}{\det g} \tag{3.31}$$

Then, from (3.29) and (3.30), we obtain

$$b^{\gamma}_{\alpha} = g^{\gamma\beta}b_{\alpha\beta} \tag{3.32}$$

which shows that b^β_α is a tensor of type (1, 1), obtained by raising up an index of the second fundamental tensor b. We call it the Weingarten tensor of S, and it has the components of (3.32), that is

$$b^1_1 = \frac{LG - MF}{\det g} \quad b^2_1 = \frac{EM - FL}{\det g} \quad b^1_2 = \frac{GM - FN}{\det g} \quad b^2_2 = \frac{EN - FM}{\det g} \tag{3.33}$$

Now it is clear that the second fundamental form of a surface is a euclidean invariant of that surface, since we defined it geometrically starting from the Gauss map.

EXERCISES

3.1 Write the equations of the tangent plane and of the normal line of a surface S, defined by an implicit equation $F(x^1, x^2, x^3) = 0$, at an arbitrary point of S.

3.2 Consider a surface defined by an explicit equation $z = f(x, y)$ ($x = x^1$, $y = x^2$, $z = x^3$). Write the equations of the tangent plane and of the normal line at an arbitrary point of the surface. (Use the Monge notation $p = \partial f/\partial x$, $q = \partial f/\partial y$.)

3.3 Consider the surface S defined by the parametric equations

$$x^1 = u^3 \sin^3 v \qquad x^2 = u^3 \cos^3 v \qquad z = (a^2 - u^2)^{\frac{1}{2}} \qquad a = \text{const}$$

and let p_1, p_2, and p_3 be the intersection point of an arbitrary tangent plane of S with the axes Ox^1, Ox^2, and Ox^3, respectively. Prove that $Op_1^2 + Op_2^2 + Op_3^2 = \text{const}$.

3.4 Consider the surface S defined by the parametric equations

$$x^1 = u^1 + u^2 \qquad x^2 = u^1 - u^2 \qquad x^3 = u^1 u^2$$

Find the unit normal vector $\underline{N}$, and the equation of the tangent plane at $u^1 = 1$, $u^2 = 2$. Discuss the nature of the parametric lines of the given parameterization.

3.5 Consider a curve $\underline{x} = \underline{\alpha}(t)$ with $\kappa \neq 0$ everywhere, and the locus of the points of radius vectors

$$\underline{x} = \underline{\alpha}(t) + \lambda\underline{\alpha}'(t) \qquad \lambda \neq 0$$

Prove that it is a surface and that the tangent planes along t = const are all equal.

3.6 Consider a curve $\underline{x} = \underline{\alpha}(s)$ with $\kappa \neq 0$ everywhere and with the natural parameter s. As usual, denote by $\underline{b}$ and $\underline{c}$ the unit principal normal and binormal vectors. Define a surface with the parametric equation

$$\underline{x} = \underline{\alpha}(s) + r[\underline{b}(s) \cos t + \underline{c}(s) \sin t] \qquad r = \text{const} \neq 0$$

It is called the tube of radius r around α. Discuss the regularity of this parameterization, and compute the unit normal vector field of the tube.

3.7 Let S be a surface and $\tilde{S} = M(S)$, where M is a motion in space. Prove that $\varphi = M \mid S: S \to \tilde{S}$ is a differentiable map, and compute $\varphi'_p: T_pS \to T_{M(p)}\tilde{S}$. Do the same if M is an affine transformation in space.

3.8 Prove that if all the normal lines of a connected surface S contain a fixed point p, S is contained in a sphere of center p. [Hint: First, prove the result on ranges of parameterizations. Then combine them to get the global result.]

3.9 Consider the surface defined by the explicit equation $z = f(x,y)$ $(x = x^1, y = x^2, z = x^3)$. Compute its first fundamental tensor and establish a formula for the area of a domain on this surface. (Use the Monge notation as in Exercise 3.2.)

3.10 Compute the first fundamental tensor for a surface defined by an implicit equation.

3.11 Show that the sphere with the center at the origin and the radius r can be covered by domains of parameterizations of the form

$$x^1 = r \sin\theta \cos\varphi \qquad x^2 = r\sin\theta\sin\varphi \qquad x^3 = r\cos\theta$$

Use these parameterizations to compute the first fundamental tensor of the sphere. Determine the curves that lie in the domain of such a parameterization, and cut the lines $\varphi = \text{const}$ at a constant angle. (These curves are called loxodromes.)

3.12 Consider the surface defined by the parameterization

$$x^1 = u^1 \cos u^2 \qquad x^2 = u^1 \sin u^2 \qquad x^3 = \ln\cos u^2 + u^1$$

Show that any two of the parametric lines $u^1 = \text{const}$ determine arcs of equal lengths on all the parametric lines $u^2 = \text{const}$.

3.13 Assume that the surface S has a parameterization such that in its domain

$$ds^2 = du^2 + (u^2 + a^2)\, dv^2 \qquad a = \text{const}$$

Compute the perimeter, the sum of the angles, and the area of the curvilinear triangle defined by the curves

$$u = \frac{av^2}{2} \qquad u = -\frac{av^2}{2} \qquad v = 1$$

3.14 Consider the surface defined by the parametric equations

$$x^1 = u^1\cos u^2 \qquad x^2 = u^1 \sin u^2 \qquad x^3 = \frac{(u^1)^2}{2}$$

(a) Show that it lies on a paraboloid of revolution.
(b) Find a curve that meets the parametric lines u^2 = const at the constant angle α and passes through the points $p(u_0^\lambda)$, $q(u_1^\lambda)$ (λ = 1, 2). Compute the value of α.
(c) Find the curves that bisect the angles between the parametric lines.

3.15 A parameterization with respect to which the first fundamental form of the surface takes the form

$$ds^2 = \lambda(u, v)(du^2 + dv^2)$$

is called an isothermic parameterization. Assume that for some parameterization $\underline{x}(u^1, u^2)$ one has $g_{12} = 0$ and $g_{11}/g_{22} = U^1(u^1)/U^2(u^2)$. Find a change of this parameterization that preserves the parametric curves and leads to an isothermic parameterization. [In this case the parametric net of $\underline{x}(u^1, u^2)$ is called an isothermal orthogonal net.]

3.16 Consider a surface with the parameterization

$$x^1 = u^1 \cos u^2 \qquad x^2 = u^1 \sin u^2 \qquad x^3 = \varphi(u^2)$$

called a right conoid. Determine the function φ such that the parametric net is an isothermal orthogonal net (Exercise 3.15). Prove that in this case the curves that bisect the angles between the parametric lines also form an isothermal orthogonal net.

3.17 Consider a surface defined by an explicit equation $z = f(x, y)$ ($x = x^1$, $y = x^2$, $z = x^3$). Compute its second fundamental tensor and its Weingarten tensor. (Use the Monge notation $p = f_x$, $q = f_y$, $r = f_{xx}$, $s = f_{xy}$, $t = f_{yy}$.)

3.18 Compute the second fundamental form of a surface defined by an implicit equation $F(x^1, x^2, x^3) = 0$.

3.19 Let S be a sphere of center p and radius r. Prove that the Weingarten transformation $\ell = \pm(1/r)$ id, according to the choice of the orientation. Use this result in order to deduce the second fundamental tensor of S. [Hint: Determine the Gauss map geometrically. Then use the last part of Exercise 3.7 to compute the differential of the Gauss map.]

3.20 Deduce analytically the results of Exercise 3.19 by means of the parameterization of a sphere given in Exercise 3.11.

3.21 Describe the region of the unit sphere covered by the image of the Gauss map of the surfaces

$$z = x^2 + y^2 \qquad x^2 + y^2 - z^2 = 1 \qquad x^2 = y^2 = \cosh^2 z$$

($x = x^1$, $y = x^2$, $z = x^3$), with a chosen orientation.

3.22 Consider the parameterized surface

$$x^1 = u^1 \quad x^2 = u^2 \quad x^3 = (u^1)^2 - (u^2)^2$$

What kind of a surface is it? Compute the Weingarten transformation of this surface at the origin.

3.23 The surface that can be covered by parameterizations of the form

$$x^1 = (a + b \cos u^1)\cos u^2 \quad x^2 = (a + b \cos u^1) \sin u^2$$

$$x^3 = b \sin u^1 \quad a > b$$

can be obtained by the rotation of the circle

$$(x^1 - a)^2 + (x^3)^2 = b^2 \quad x^2 = 0$$

about the x^3 axis and is called a torus. Compute the fundamental form and Weingarten transformation of the torus.

3.24 Compute the second fundamental tensor and the Weingarten tensor of the right conoid

$$x^1 = u^1 \cos u^2 \quad x^2 = u^1 \sin u^2 \quad x^3 = \varphi(u^2)$$

3.2 GEOMETRY OF THE SECOND FUNDAMENTAL FORM

This section continues with the discussion of the geometry related to the second fundamental tensor of a surface. We begin with a geometric interpretation of the tensor itself.

Let us fix a point $p_0 \in S$. We should like to discuss the geometric shape of the surface in a neighborhood of p_0, particularly the position of the neighboring points $p \in S$ with respect to the tangent plane $\pi_{p_0} S$ of S at p_0. To do this, we compute the projection of the vector $\underline{p_0 p}$ on the normal $\underline{N}$ (Fig. 3.2).

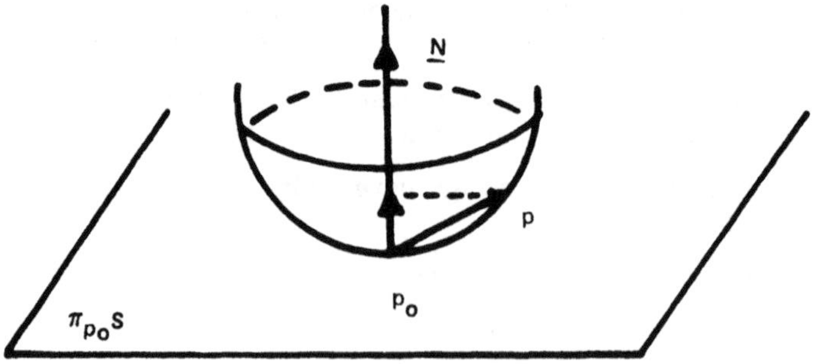

Figure 3.2

Using the Taylor formula, we get

$$\underline{p_0 p} = \underline{x}(u^\alpha) - \underline{x}(u_0^\alpha) = (\underline{x}_\alpha)_0 (u^\alpha - u_0^\alpha) + \frac{1}{2}(\underline{x}_{\alpha\beta})_0 (u^\alpha - u_0^\alpha)(u^\beta - u_0^\beta) + \cdots$$

and then

$$\underline{N} \cdot \underline{p_0 p} = \frac{1}{2} b_{\alpha\beta}(u_0^\alpha)(u^\alpha - u_0^\alpha)(u^\beta - u_0^\beta) + \cdots \tag{3.34}$$

If this quantity $\underline{N} \cdot \underline{p_0 p}$ is > 0, p lies on the same part of $\pi_{p_0} S$ as $\underline{N}$; if it is < 0, p is on the opposite side. In this sense, we can say that the second fundamental form is the main part of the product $\underline{N} \cdot \underline{p_0 p}$, and that it determines the geometric shape of S about p_0.

To be more precise, let us examine it in the following discussion. If all the $b_{\alpha\beta}(u_0^\alpha) = 0$, we shall say that p_0 is a planar point, and the sign of $\underline{N} \cdot \underline{p_0 p}$ will be defined by terms of higher order in the Taylor development (3.34). The name comes from

PROPOSITION 3.3 If U is an open connected subset of a surface S and consists of planar points only, then U belongs to a plane.

Proof: First, suppose that U belongs to the range of a parameterization $\underline{x}$. Then $b_{\alpha\beta}(u^\alpha) = 0$ means $\underline{N}_\alpha \cdot \underline{x}_\beta = 0$. Moreover, we also have $\underline{N}_\alpha \cdot \underline{N} = 0$, which follows by differentiating $\underline{N}^2 = 1$. Since $\underline{x}_1$, $\underline{x}_2$, $\underline{N}$ are independent vectors, it follows that $\underline{N}_\alpha = \underline{0}$ $(\alpha = 1, 2)$, and therefore $\underline{N}$ is a constant vector. Obviously, this constant vector satisfies $\underline{N} \cdot d\underline{x} = 0$, whence $\underline{N} \cdot \underline{x} = \text{const}$, which is the equation of some plane in E^3.

Now, if U is not in the range of a parameterization, we can express it in the form $U = \cup_\sigma U_\sigma$, where each U_σ is connected and in the range of a parameterization, and according to the proof above, each U_σ will be in a plane. Since S is a differentiable surface, these planes must coincide. Q.E.D.

Now, if p_0 is not a planar point, the sign of (3.34) is determined by the given term of the Taylor development, and by elementary mathematics, three cases can occur:

(a) $b_{11}b_{22} - b_{12}^2 = \det b > 0$. In this case $\underline{N} \cdot \underline{p_0 p}$ has a fixed sign and a whole neighborhood of p_0 on S lies on the same side of the tangent plane $\pi_{p_0} S$. Such a point p_0 is called elliptic, and the respective neighborhood of S looks like that in Fig. 3.2.

(b) $b_{11}b_{22} - b_{12}^2 = \det b < 0$. In this case the sign of $\underline{N} \cdot \underline{p_0 p}$ is not fixed, and near p_0 the surface has points on the two sides of the tangent plane $\pi_{p_0} S$. S has then the shape of a saddle (Fig. 3.3), and p_0 is called a hyperbolic point.

(c) $b_{11}b_{22} - b_{12}^2 = \det b = 0$. In this case the sign of $\underline{N} \cdot \underline{p_0 p}$ is fixed except for some curve on S, where the first term of the development (3.34)

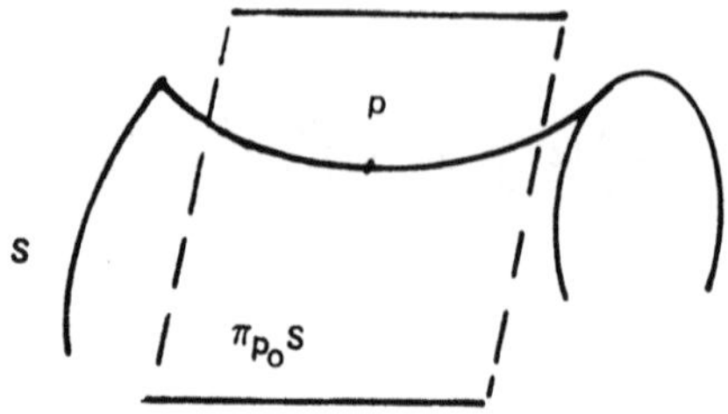

Figure 3.3

vanishes. The surface looks like the one shown in Fig. 3.4 and p_0 is called a parabolic point.

Let us now present some more geometrical notions.

DEFINITION 3.2 Two tangent vectors $\underline{v}$, $\underline{w} \in T_pS$ are called conjugate if $b(\underline{v}, \underline{w}) = 0$. The directions of such vectors are called conjugate directions. The same term can be used for vector fields that satisfy the condition above at every point. The integral paths of two conjugate vector fields define a conjugate net.

For instance, the parametric net of a parameterization is conjugate iff $b_{12} = 0$.

DEFINITION 3.3 A self-conjugate vector (direction) is called an asymptotic vector (direction). The integral paths of an asymptotic field are called asymptotic lines, and they form the asymptotic net of the surface.

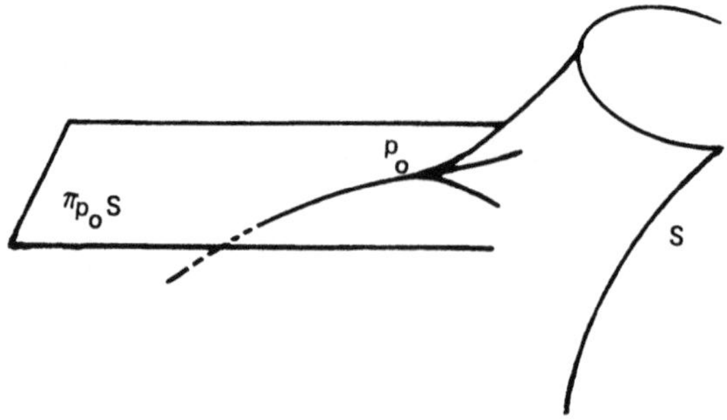

Figure 3.4

It follows that the asymptotic vectors are defined by the equation

$$b(\underline{v}, \underline{v}) = b_{\alpha\beta}\xi^{\alpha}\xi^{\beta} = 0 \tag{3.35}$$

and we see that we have two distinguished (real) asymptotic directions at every hyperbolic point, one asymptotic direction (a "double" one) at every parabolic point, and no real asymptotic directions ("two imaginary conjugate directions") at every elliptic point of S. At the planar points, we have "indeterminate" asymptotic directions.

The asymptotic net is defined by the differential equation

$$b_{\alpha\beta} \frac{du^{\alpha}}{dt} \frac{du^{\beta}}{dt} = 0 \tag{3.36}$$

or simply

$$\Psi = b_{\alpha\beta}\, du^{\alpha}\, du^{\beta} = 0 \tag{3.37}$$

where the differentials are along the looked-for paths.

Let us note that (3.37) can be written down along any curve on S, and this is the classical manner of describing the second fundamental form of a surface. If the expressions of $b_{\alpha\beta}$ using a parameterization $\underline{x}$ are replaced in (3.37) this formula becomes

$$\Psi = -d\underline{N} \cdot d\underline{x} = \underline{N} \cdot d^2\underline{x} \tag{3.38}$$

It is worthwhile to show how to find the local solutions of an equation of the form (3.37) if its coefficients do not vanish simultaneously. If $b_{11} = b_{22} = 0$, $b_{12} \neq 0$, the equation becomes $du^1\, du^2 = 0$, and we have two solutions, $u^1 = \text{const}$ and $u^2 = \text{const}$, respectively. (That is, $b_{11} = b_{22} = 0$ characterizes parameterizations whose parametric net is the asymptotic net. Note that in a neighborhood of a hyperbolic point such a parameterization is always available by Proposition 3.1.)

Now, if $b_{11} \neq 0$ for example, we can express (3.37) in the equivalent form

$$b_{11}\left(\frac{du^1}{du^2}\right)^2 + 2b_{12}\frac{du^1}{du^2} + b_{22} = 0$$

which has two solutions

$$\frac{du^1}{du^2} = f_1(u^1, u^2) \qquad \frac{du^1}{du^2} = f_2(u^1, u^2)$$

Thereby we obtain two usual ordinary differential equations, whose solutions are the solutions of (3.37).

It is obvious that the asymptotic net has an invariant geometrical meaning, since b is a euclidean invariant. This also follows from

PROPOSITION 3.4 A curve γ on the surface S is an asymptotic line iff its osculating plane coincides with the tangent plane of S at every point of the curve.

Proof: Take $p \in \gamma \subset S$ and a parameterization $\underline{x}$ of S about p. Then the osculating plane of γ at p is defined by the vectors $d\underline{x}$ and $d^2\underline{x}$, and since $\underline{N} \cdot d\underline{x} = 0$, this plane is $\pi_{p_0} S$ iff $\underline{N} \cdot d^2\underline{x} = 0$, that is $\Psi = 0$. Q.E.D.

We proceed with

DEFINITION 3.4 A vector $\underline{v} \in T_pM$, $\underline{v} \neq \underline{0}$, is called a <u>principal vector</u>, and its direction is called a <u>principal direction</u>, if $\underline{v}$ is an eigenvector of the Weingarten transformation ℓ. That is, some scalar κ exists such that

$$\ell(v) = \kappa v \tag{3.39}$$

The corresponding eigenvalues κ of ℓ are called the <u>principal curvatures</u>.

Tor a discussion of these notions, let us note that the componentwise form of Eq. (3.39) is

$$(b^\alpha_\beta - \kappa\delta^\alpha_\beta)\xi^\beta = 0 \tag{3.40}$$

where δ is the Kronecker index. By contracting with $g_{\gamma\alpha}$, we get the equivalent form

$$(b_{\gamma\beta} - \kappa g_{\gamma\beta})\xi^\beta = 0 \tag{3.41}$$

which is a linear homogeneous system whose nontrivial solutions define the principal vectors.

Such solutions exist iff

$$\det(b_{\gamma\beta} - \kappa g_{\gamma\beta}) = \begin{vmatrix} b_{11} - \kappa g_{11} & b_{12} - \kappa g_{12} \\ b_{21} - \kappa g_{21} & b_{22} - \kappa g_{22} \end{vmatrix} = 0 \tag{3.42}$$

which defines, therefore, the principal curvatures. Since (3.42) is of the second degree in κ, we can generally speak of two such curvatures κ_1, κ_2.

DEFINITION 3.5

(a) $H = \kappa_1 + \kappa_2$ is called the <u>mean curvature</u> of S (at p).

(b) $K = \kappa_1\kappa_2$ is called the <u>total</u> or <u>Gaussian curvature</u> of S (at p).

Note that κ_1, κ_2, H, K are geometric invariants since the Weingarten transformation ℓ is such.

It follows from Definition 3.5 that (3.42) and the equivalent equation $\det(b^\alpha_\beta - \kappa\delta^\alpha_\beta) = 0$ obtained from (3.40) have the form

$$\kappa^2 - H\kappa + K = 0 \tag{3.43}$$

whence, by comparison,

$$H = g^{\alpha\beta}b_{\alpha\beta} = \operatorname{tr} \ell \quad (\operatorname{tr} = \text{trace})$$
$$K = \frac{\det (b_{\alpha\beta})}{\det (g_{\alpha\beta})} = \det \ell \tag{3.44}$$

These are very important invariants of a surface. For the moment, let us remark only that the sign of K and of det $(b_{\alpha\beta})$ is the same, and therefore an elliptic point is characterized by $K > 0$, a hyperbolic point by $K < 0$, and a parabolic point by $K = 0$. (At a parabolic point one of the principal curvatures vanishes.)

Let us deduce some more results about the principal curvatures.

PROPOSITION 3.5 The principal curvatures of a surface S are always real.

Proof: This is a well-known algebraic property since ℓ is a self-adjoint operator (Proposition 3.2), but we can also give here a straightforward proof. Fix a point p, and solve (3.43) at that point. We obtain

$$\kappa_{1,2} = \frac{H \pm (H^2 - 4K)^{\frac{1}{2}}}{2} \tag{3.45}$$

Now let us construct two orthogonal vector fields in a neighborhood of p, and take a parameterization whose parametric net consists of integral paths of these fields (Proposition 3.1). Then $g_{12} = 0$ and we get

$$H^2 - 4K = \frac{(g_{11}b_{22} - g_{22}b_{11})^2 + 4g_{11}b_{12}^2}{(\det g)^2} \tag{3.46}$$

which is nonnegative, since obviously, $g_{11} > 0$, $g_{22} > 0$. Q.E.D.

DEFINITION 3.6 If $\kappa_1 = \kappa_2$ at a point $p \in S$, p is called an <u>umbilical point (umbilic)</u>.

It follows from (3.45) that the umbilics are defined by the equation $H^2 = 4K$, and hence they form a closed subset $\mathcal{O}$ of S. Moreover, κ_1, κ_2 are always continuous but they are differentiable on $S \setminus \mathcal{O}$ only, while K and H are always differentiable.

Formula (3.46) yields another form of the umbilicity condition, namely, that $g_{12} = 0$ implies $b_{12} = 0$ and $g_{11}b_{22} = g_{22}b_{11}$. This is clearly equivalent to the invariant condition

$$b_{\alpha\beta} = \rho g_{\alpha\beta} \tag{3.47}$$

for some factor ρ. In other words, p is an umbilic iff the two fundamental forms are proportional at p. Or, by raising the index β, $\ell = \rho \cdot \mathrm{id}$. In particular, a planar point is an umbilic.

It follows that an open subset of a plane is a surface consisting of umbilics only. It is also simple to see that the same holds for an open subset of a sphere. Indeed, if R is the radius of the sphere and if the normal is oriented toward the center of the sphere, we have $\underline{N} = -(1/R)\underline{x}$, and by a differentiation we see that the Weingarten transformation is $\ell = (1/R)\mathrm{id}$. Q.E.D.

The converse is also true, that is, we have

PROPOSITION 3.6 Let S be an orientable connected surface all of whose points are umbilics. Then S is contained either in a plane or in a sphere. If, moreover, S is closed in E^3, it is either a plane or a sphere.

Proof: First, let us study a connected neighborhood of S contained in the range of a parameterization. The condition $\ell = \rho$ id means that $b_\alpha^\beta = \rho\delta_\alpha^\beta$, and the Weingarten formulas (3.20) yield

$$\underline{N}_\alpha = -\rho\underline{x}_\alpha \tag{3.48}$$

On the other hand, (3.47) can be written as

$$\underline{N} \cdot \underline{x}_{\alpha\beta} = \rho\underline{x}_\alpha \cdot \underline{x}_\beta$$

whence, by differentiating with respect to u^γ, we get

$$\underline{N}_\gamma \cdot \underline{x}_{\alpha\beta} + \underline{N} \cdot \underline{x}_{\alpha\beta\gamma} = \rho_\gamma\underline{x}_\alpha \cdot \underline{x}_\beta + \rho\underline{x}_{\alpha\gamma} \cdot \underline{x}_\beta + \rho\underline{x}_\alpha \cdot \underline{x}_{\beta\gamma}$$

If we replace $\underline{N}_\gamma$ by (3.48) here, we deduce that $\rho_\gamma\underline{x}_\alpha \cdot \underline{x}_\beta = \rho_\gamma g_{\alpha\beta}$ is symmetric in all its indices. Therefore, we have

$$\rho_\gamma g_{\alpha\beta} = \rho_\alpha g_{\gamma\beta}$$

and if we contract here with $g^{\alpha\beta}$, we get $2\rho_\gamma = \rho_\gamma$, which implies that $\rho_\gamma = 0$. In view of the connectedness condition, this means that $\rho = \mathrm{const}$.

Now, if $\rho = 0$ all the points are planar, and they belong to a plane (Proposition 3.3). If $\rho \neq 0$, (3.48) gives

$$d\underline{x} + \frac{1}{\rho}d\underline{N} = 0$$

whence

$$\underline{x} + \frac{1}{\rho}\underline{N} = \underline{a}$$

where $\underline{a}$ is a constant vector. Therefore, we have

$$(\underline{x} - \underline{a})^2 = \frac{1}{\rho^2}$$

which shows that the neighborhood of S studied lies on a sphere.

Finally, we see that the same is true for the whole of S, by dividing S into pieces as in the case above, and by relying on the differentiability of S (as in the proof of Proposition 3.3). The final assertion of Proposition 3.6 is then obvious. Q.E.D.

Now let us discuss the principal vectors. We already know that these vectors are defined by (3.40) or (3.41), with κ replaced by κ_1, κ_2. First, we note the following result (which is well known for self-adjoint operators):

PROPOSITION 3.7 Two principal vectors that correspond to distinct principal curvatures are orthogonal.

Proof: If $\ell(\underline{v}_1) = \kappa_1\underline{v}_1$, $\ell(\underline{v}_2) = \kappa_2\underline{v}_2$, $\kappa_1 \neq \kappa_2$, we get from the self-adjointness of ℓ:

$$0 = \ell(\underline{v}_1) \cdot \underline{v}_2 - \underline{v}_1 \cdot \ell(\underline{v}_2) = (\kappa_1 - \kappa_2)\underline{v}_1 \cdot \underline{v}_2$$

and since $\kappa_1 \neq \kappa_2$, $\underline{v}_1 \cdot \underline{v}_2 = 0$. Q.E.D.

It follows that if p is a nonumbilical point of S, every principal curvature defines one and only one principal direction, since the tangent space of S is two-dimensional. Therefore, through every such p there are exactly two principal directions orthogonal to one another.

If p is an umbilic, we see by (3.41) and (3.47) that every direction in T_pS is a principal direction and, conversely, this property characterizes umbilics. (At an umbilic, the principal directions are "indeterminate.")

Another interesting result is

PROPOSITION 3.8 Every principal direction is conjugate to its orthogonal direction.

Proof: Indeed, if $\ell(\underline{v}) = \kappa\underline{v}$ and $\underline{v} \cdot \underline{w} = 0$, we have $b(\underline{v}, \underline{w}) = \ell(\underline{v}) \cdot \underline{w} = \kappa\underline{v} \cdot \underline{w} = 0$. Q.E.D.

In particular, if $p \in S$ is not an umbilic, the principal directions at p are conjugate directions.

PROPOSITION 3.9 $\underline{v} \neq \underline{0}$ is a principal vector iff it satisfies the condition

$$(\underline{N}, \ell\underline{v}, \underline{v}) = 0 \tag{3.49}$$

Proof: If $\underline{v}$ is principal, $\ell\underline{v}$ is collinear with $\underline{v}$, and (3.49) holds. Conversely, if (3.49) holds, we get $\ell\underline{v} = \alpha\underline{v} + \beta\underline{N}$, and, since $\ell\underline{v} \in T_pS$, $\beta = 0$. Hence $\underline{v}$ is principal. Q.E.D.

This proposition suggests defining a new two-times covariant symmetric tensor c by

$$c(\underline{v},\underline{w}) = -\frac{(\det g)^{\frac{1}{2}}}{2}[(\underline{N},\ell\underline{v},\underline{w}) + (\underline{N},\ell\underline{w},\underline{v})] \tag{3.50}$$

Then we get for the principal directions the equation

$$c(\underline{v},\underline{v}) = c_{\alpha\beta}\xi^{\alpha}\xi^{\beta} = 0 \tag{3.51}$$

where $c_{\alpha\beta} = c(\underline{x}_{\alpha},\underline{x}_{\beta})$, and by the Weingarten formulas, we get

$$c_{11} = g_{11}b_{12} - g_{12}b_{11} \qquad c_{12} = c_{21} = \frac{1}{2}(g_{11}b_{22} - g_{22}b_{11})$$

$$c_{22} = g_{12}b_{22} - g_{22}b_{12} \tag{3.52}$$

This shows that Eq. (3.51) can be put in the form

$$c(\underline{v},\underline{v}) = \begin{vmatrix} (\xi^2)^2 & -\xi^1\xi^2 & (\xi^1)^2 \\ g_{11} & g_{12} & g_{22} \\ b_{11} & b_{12} & b_{22} \end{vmatrix}$$

$$= \begin{vmatrix} g_{11}\xi^1 + g_{12}\xi^2 & g_{21}\xi^1 + g_{22}\xi^2 \\ b_{11}\xi^1 + b_{12}\xi^2 & b_{21}\xi^1 + b_{22}\xi^2 \end{vmatrix} = \frac{1}{4}\frac{\partial(g,b)}{\partial(\xi^1,\xi^2)} = 0 \tag{3.53}$$

Because of the last expression, we shall also denote c by J(g,b) and call it the jacobian tensor of the two first fundamental tensors of the surface S.

Another important notion is introduced by

DEFINITION 3.7 The integral paths of the principal vector fields are called the lines of curvature of the surface, and they define the curvature net.

The properties of the curvature net are deduced from the discussion above. For instance, it is a net that is simultaneously orthogonal and conjugate (we are referring here to nonumbilics). The equation of the lines of curvature follows by setting $\xi^1 = du^1$, $\xi^2 = du^2$ in (3.53), and it is

$$c_{\alpha\beta}du^{\alpha}du^{\beta} = \begin{vmatrix} (du^2)^2 & -du^1du^2 & (du^1)^2 \\ g_{11} & g_{12} & g_{22} \\ b_{11} & b_{12} & b_{22} \end{vmatrix} = 0 \tag{3.54}$$

The integration of (3.54) can be performed as discussed for (3.37).

In a neighborhood of a nonumbilical point, we can get, by Proposition 3.1, a parameterization whose parametric net will be the curvature net of the surface. This is very convenient since, because this net is both orthogonal and conjugate, we have $g_{12} = 0$, $b_{12} = 0$, which in turn can simplify many computations. Let us note again that by the geometric character of the Weingarten transformation ℓ, the principal directions and the curvature lines are geometric invariants of a surface.

We end this section by sketching the Gauss interpretation of the total curvature K, which is very interesting from both the historical and mathematical viewpoints. Let us assume that p_0 is an elliptic point of the surface S. Take a parameterized neighborhood U of p_0 on S and its spherical image by the Gauss map γ, $\gamma(U)$. Then $\gamma(U)$ has the parameterization $\underline{x} = \underline{N}(u^{\alpha})$. Indeed, by using the Weingarten formulas we get

$$\underline{N}_1 \times \underline{N}_2 = (b_1^{\alpha} \underline{x}_{\alpha}) \times (b_2^{\beta} \underline{x}_{\beta}) = K\underline{x}_1 \times \underline{x}_2 \neq \underline{0}$$

Moreover, since $K > 0$ (p is elliptic), we have

$$(\det \tilde{g})^{\frac{1}{2}} = |\underline{N}_1 \times \underline{N}_2| = K(\det t)^{\frac{1}{2}}$$

where $\tilde{g}$ is the first fundamental tensor of $\gamma(U)$.

It follows that the area A of U and the area $\mathscr{A}$ of $\gamma(U)$ are related by

$$\mathscr{A} = \iint_{\underline{x}^{-1}(U)} K(\det g)^{\frac{1}{2}}\, du^1\, du^2 = K(\tilde{u}^1, \tilde{u}^2)A$$

where $(\tilde{u}^1, \tilde{u}^2)$ is some "intermediate" point in U. (We are using here the integral mean value theorem.) This relation yields

$$K(p_0) = \lim_{U \to p_0} \frac{\mathscr{A}}{A}$$

which is the announced Gauss interpretation.

EXERCISES

3.25 Determine which of the points of the surface $z = x^3 + y^3$ ($x = x^1$, $y = x^2$, $z = x^3$) are planar, elliptic, hyperbolic, and parabolic.

3.26 Classify the points of the various quadric surfaces.

3.27 Show that if the tangent plane of a surface S along a curve C is constant, the points of C are either planar or parabolic points of C. [Hint: Use either the geometric interpretation of the second fundamental form or explicit equations for S and C.]

3.28 Consider a surface defined by a parameterization of the form $\underline{x} = \underline{a}(u^1) + \underline{b}(u^2)$, with the parameters u^1, u^2. Prove that the parametric net is conjugate. (Such a surface is called a translation surface.)

3.29 Consider the loxodromes of the sphere of Exercise 3.11, which cut the lines $\varphi = \text{const}$ at the constant angle α. Find the curves of the sphere that are conjugate to those loxodromes.

3.30 Find the asymptotic lines of the various quadric surfaces.

3.31 Determine the asymptotic lines of the surface $x^3 = x^2 \sin x^1$ and their orthogonal trajectories.

3.32 Find the asymptotic lines and the principal curvatures of the Enneper surface,

$$x^1 = u^1 - \frac{(u^1)^3}{3} + u^1(u^2)^2 \qquad x^2 = u^2 - \frac{(u^2)^3}{3} + u^2(u^1)^2 \qquad x^3 = (u^1)^2 - (u^2)^2$$

3.33 Prove that if $p \in S$ is a nonplanar point, and if the mean curvature of S at p vanishes, the asymptotic directions of S at p are orthogonal.

3.34 Consider the surface defined by the parameterization

$$x^1 = u \cos v \qquad x^2 = u \sin v \qquad x^3 = a \ln \frac{a + (a^2 - u^2)^{\frac{1}{2}}}{u} - (a^2 - u^2)^{\frac{1}{2}}$$

where $a = \text{const}$. Prove that this surface has a constant negative total curvature equal to $-1/a^2$. Find the asymptotic lines of the surface. (This surface is called Beltrami's pseudosphere.)

3.35 Consider the surface

$$x^1 = a \cosh u^1 \cosh u^2 \qquad x^2 = b \sinh u^1 \cosh u^2 \qquad x^3 = c \sinh u^2$$

What kind of surface is it? Compute the principal curvatures at the intersection points of the surface with the coordinate axes.

3.36 Prove that the mean curvature of the surface

$$x^1 = u \cos v \qquad x^2 = u \sin v \qquad x^3 = v$$

vanishes.

3.37 Find the lines of curvature of the surface defined in Exercise 3.36.

3.38 Let $p \in S$ be a hyperbolic point. Prove that the principal directions bisect the asymptotic directions at p.

3.39 Using the lines of curvature of a surface as parametric lines, establish a formula for the angle between two conjugate directions.

3.40 Consider the surface S with

$$x^1 = au + \sin u \cosh v \qquad x^2 = v + a \cos u \sinh v$$

$$x^3 = (1 - a^2)^{\frac{1}{2}} \cos u \cosh v$$

Prove that the mean curvature of S vanishes, and find its asymptotic and curvature lines.

3.41 Find the asymptotic lines and the curvature lines of the surface

$$x^1 = \frac{a}{2}(u^1 + u^2) \qquad x^2 = \frac{b}{2}(u^1 - u^2) \qquad x^3 = \frac{u^1 u^2}{2}$$

3.42 Prove that the definition of the total curvature is meaningful for nonorientable surfaces as well, but that this is not true for the mean curvature. [Hint: Study the behavior of the Weingarten transformation under a change of the normal orientation.]

3.3 THE COVARIANT DERIVATIVE AND THE FUNDAMENTAL THEOREM

As in the theory of curves (see Theorem 2.2), we want to show that the fundamental tensors introduced in Sec. 3.2 define the surface up to a motion in space. To do this we shall use a method similar to that used in Theorem 2.2, but instead of the Frenet frame of a curve, we consider the affine frame defined by $(\underline{x}_1, \underline{x}_2, \underline{N})$ at each point p of the surface S. This is the moving frame method, and the frame $(\underline{x}_1, \underline{x}_2, \underline{N})$ is called the Darboux frame of S (at p).

The case of a surface is essentially different from the case of a curve since we have two independent variables. Henceforth, some rather lengthy preparation is needed. However, this is not a waste of time since it gives important information on the geometry of a surface.

We begin by discussing the covariant derivative.

Let us consider the surface $S \subset E^3$, a point $p \in S$, and a tangent vector $\underline{v} \in T_pS$. Now, assume that $\underline{w}$ is a tangent vector field defined on an open neighborhood of p. Then the action $\underline{v}(\underline{w})$ is defined by (1.75), and the result is a vector in E^3. Let us decompose this vector by

$$\underline{v}(\underline{w}) = D_{\underline{v}}\underline{w} + \alpha\underline{N} \tag{3.55}$$

where $D_{\underline{v}}\underline{w} \in T_pS$.

First, we note that

$$\alpha = \underline{N} \cdot \underline{v}(\underline{w}) = -\underline{v}(\underline{N}) \cdot \underline{w} = b(\underline{v}, \underline{w}) \tag{3.56}$$

where the second equality follows by applying $\underline{v}$ to $\underline{N} \cdot \underline{w} = 0$ and using (1.79), and the last equality follows by (3.25). Hence (3.55) becomes

$$\underline{v}(\underline{w}) = D_{\underline{v}}\underline{w} + b(\underline{v}, \underline{w})\underline{N} \tag{3.57}$$

and this can be extended to vector fields, $\underline{v}$, $\underline{w}$. Equation (3.57) is called the Gauss equation.

$D_{\underline{v}}$ is an operator acting on fields $\underline{w}$. It can be considered as being defined by the Gauss equation (3.57) and it sends $\underline{w}$ to a tangent vector $D_{\underline{v}}\underline{w} \in T_pS$.

DEFINITION 3.8 $D_{\underline{v}}$ is called the covariant derivative with respect to the vector $\underline{v}$.

The main properties of the covariant derivative are given by

PROPOSITION 3.10 The operator $D_{\underline{v}}$ satisfies the relations

(a) $D_{\underline{v}}(\alpha\underline{w}_1 + \beta\underline{w}_2) = \alpha D_{\underline{v}}\underline{w}_1 + \beta D_{\underline{v}}\underline{w}_2$

(b) $D_{\alpha\underline{v}_1+\beta\underline{v}_2}\underline{w} = \alpha D_{\underline{v}_1}\underline{w} + \beta D_{\underline{v}_2}\underline{w}$

(c) $D_{\underline{v}}(f\underline{w}) = \underline{v}(f)\underline{w} + fD_{\underline{v}}\underline{w}$

(d) $D_{f\underline{v}}\underline{w} = fD_{\underline{v}}\underline{w}$

(e) $\underline{v}(\underline{w}_1 \cdot \underline{w}_2) = (D_{\underline{v}}\underline{w}_1) \cdot \underline{w}_2 + \underline{w}_1 \cdot (D_{\underline{v}}\underline{w}_2)$

(f) $D_{\underline{v}}\underline{w} - D_{\underline{w}}\underline{v} = [\underline{v}, \underline{w}]$

where $\underline{v}$, $\underline{w}$ are vector fields on S, α, $\beta \in \mathbb{R}$, and f is a differentiable function on S.

Proof: Relations (a), (b), (c), and (d) follow immediately if we replace $D_{\underline{v}}$ from (3.57), and use formulas (1.76)-(1.78) together with the bilinearity of b. Relation (e) follows from (1.79) and (3.57) since the scalar product of $\underline{N}$ with any tangent vector is zero. (This property is known as the Ricci lemma.) Finally, relation (f) (where [,] denotes the bracket of Sec. 1.6) follows by (3.57) and (1.80), in view of the symmetry of b. Q.E.D.

Unlike the usual derivatives, the higher-order covariant derivatives are not symmetric, which leads to the definition of the following important operator:

$$R(\underline{v}, \underline{w}) = D_{\underline{v}}D_{\underline{w}} - D_{\underline{w}}D_{\underline{v}} - D_{[\underline{v}, \underline{w}]} \tag{3.58}$$

which is called the curvature of D. It associates with every vector field $\underline{u}$ the vector field $R(\underline{v}, \underline{w})\underline{u}$.

PROPOSITION 3.11 For every differentiable function f, we have

$$R(f\underline{v},\underline{w})\underline{u} = R(\underline{v},f\underline{w})\underline{u} = R(\underline{v},\underline{w})(f\underline{u}) = fR(\underline{v},\underline{w})\underline{u}$$

Proof: Compute using (3.58) and Proposition 3.10.

We now prove the following important result:

THEOREM 3.1 The covariant derivative and its curvature depend only on the first fundamental form.

Proof: If $\underline{v} = \xi^\alpha \underline{x}_\alpha$, $\underline{w} = \eta^\beta \underline{x}_\beta$ (locally, and with respect to a parameterization $\underline{x}$), we have by Proposition 3.10,

$$D_{\underline{v}}\underline{w} = \xi^\alpha\left(\frac{\partial \eta^\beta}{\partial u^\alpha}\underline{x}_\beta + \eta^\beta D_{\underline{x}_\alpha}\underline{x}_\beta\right)$$

whence, by defining the coefficients $\Gamma^\sigma_{\alpha\beta}$ by

$$D_{\underline{x}_\beta}\underline{x}_\alpha = \Gamma^\sigma_{\alpha\beta}\underline{x}_\sigma \tag{3.59}$$

we get

$$D_{\underline{v}}\underline{w} = \xi^\alpha\left(\frac{\partial \eta^\sigma}{\partial u^\alpha} + \Gamma^\sigma_{\beta\alpha}\eta^\beta\right)\underline{x}_\sigma = \xi^\alpha \eta^\sigma{}_{/\alpha}\underline{x}_\sigma \tag{3.60}$$

where we set

$$\eta^\sigma{}_{/\alpha} = \frac{\partial \eta^\sigma}{\partial u^\alpha} + \Gamma^\sigma_{\beta\alpha}\eta^\beta \tag{3.61}$$

Hence the covariant derivative is determined by the coefficients $\Gamma^\sigma_{\alpha\beta}$. They are called the Christoffel symbols of the second kind.

To compute these coefficients, we take three fields $\underline{u}$, $\underline{v}$, $\underline{w}$, and use the Ricci lemma:

$$\underline{u}(\underline{v}\cdot\underline{w}) = D_{\underline{u}}\underline{v}\cdot\underline{w} + \underline{v}\cdot D_{\underline{u}}\underline{w}$$

$$-\underline{v}(\underline{w}\cdot\underline{u}) = -D_{\underline{v}}\underline{w}\cdot\underline{u} - \underline{w}\cdot D_{\underline{v}}\underline{u}$$

$$\underline{w}(\underline{u}\cdot\underline{v}) = D_{\underline{w}}\underline{u}\cdot\underline{v} + \underline{u}\cdot D_{\underline{w}}\underline{v}$$

By adding these relations and using Proposition 3.10(f), we obtain

$$2\underline{v}\cdot D_{\underline{w}}\underline{u} = \underline{u}(\underline{w}\cdot\underline{v}) - u\cdot[\underline{w},\underline{v}] - \underline{v}(\underline{w}\cdot\underline{u}) + \underline{v}\cdot[\underline{w},\underline{u}] + \underline{w}(\underline{u}\cdot\underline{v}) - \underline{w}\cdot[\underline{u},\underline{v}]$$

Now take $\underline{u} = \underline{x}_\alpha$, $\underline{v} = \underline{x}_\beta$, $\underline{w} = \underline{x}_\gamma$. Then the relation above becomes

$$2g_{\beta\sigma}\Gamma^{\sigma}_{\alpha\gamma} = \frac{\partial g_{\beta\gamma}}{\partial u^{\alpha}} - \frac{\partial g_{\alpha\gamma}}{\partial u^{\beta}} + \frac{\partial g_{\alpha\beta}}{\partial u^{\gamma}} = [\alpha,\gamma;\beta] \tag{3.62}$$

where the last bracket is a new notation, and $[\alpha,\gamma;\beta]$ are called the Christoffel symbols of the first kind. Finally, by contracting in (3.62) by $g^{\lambda\beta}$, we obtain

$$\Gamma^{\lambda}_{\alpha\gamma} = g^{\lambda\beta}[\alpha,\gamma;\beta] \tag{3.63}$$

which proves that the covariant derivative is completely defined by the first fundamental form.

Then it is obvious that the same is true for its curvature. More specifically, it follows from (3.58) and Proposition 3.11 that R is completely defined by the coefficients of

$$R(\underline{x}_{\alpha},\underline{x}_{\beta})\underline{x}_{\gamma} = R^{\lambda}{}_{\gamma\beta\alpha}\underline{x}_{\lambda} \tag{3.64}$$

which are easily seen to be given by

$$R^{\lambda}{}_{\gamma\beta\alpha} = \Gamma^{\lambda}_{\gamma\beta,\alpha} - \Gamma^{\lambda}_{\gamma\alpha,\beta} + \Gamma^{\rho}_{\gamma\beta}\Gamma^{\lambda}_{\rho\alpha} - \Gamma^{\rho}_{\gamma\alpha}\Gamma^{\lambda}_{\rho\beta} \tag{3.65}$$

REMARK By using the transformation law for $\underline{x}_{\alpha}$, one can see that the $\Gamma^{\alpha}_{\beta\gamma}$ are not the components of a tensor but that, in view of (3.64), $R^{\lambda}{}_{\gamma\beta\alpha}$ are the components of a tensor of the type (1,3), called the curvature tensor of S. The $\eta^{\sigma}{}_{/\alpha}$ of (3.61) are also the components of a tensor, of type (1,1), and classical books on differential geometry call this the covariant derivative.

Furthermore, the Gauss equation (3.57) allows us to derive some important relations between the covariant derivative (i.e., the first fundamental form) and the second fundamental form of a surface.

THEOREM 3.2 For any three tangent vector fields $\underline{u}$, $\underline{v}$, $\underline{w}$ of S, the following relations hold:

$$R(\underline{u},\underline{v})\underline{w} = b(\underline{v},\underline{w})\ell(\underline{u}) - b(\underline{u},\underline{w})\ell(\underline{v}) \tag{3.66}$$

$$D_{\underline{u}}(\ell(\underline{v})) - D_{\underline{v}}(\ell(\underline{u})) - \ell([\underline{u},\underline{v}]) = 0 \tag{3.67}$$

Proof: Apply $\underline{u}$ to both sides of the Gauss equation (3.57), again use the Gauss equation for $\underline{u}(D_{\underline{v}}\underline{w})$, and take into account formulas (3.23) and (3.26). This yields

$$\underline{u}\,\underline{v}(\underline{w}) = D_{\underline{u}}D_{\underline{v}}\underline{w} - b(\underline{v},\underline{w})\ell(\underline{u}) + [\underline{u}(\ell(\underline{v})\cdot\underline{w}) + \ell(\underline{u})\cdot D_{\underline{v}}\underline{w}]\underline{N}$$

Similarly,

$$\underline{v}\,\underline{u}(\underline{w}) = D_{\underline{v}} D_{\underline{u}} \underline{w} - b(\underline{u}, \underline{w})\ell(\underline{v}) + [\underline{v}(\ell(\underline{u}) \cdot \underline{w}) + \ell(\underline{v}) \cdot D_{\underline{u}} \underline{w}]\underline{N}$$

and

$$[\underline{u}, \underline{v}]\,(\underline{w}) = D_{[\underline{u}, \underline{N}]} \underline{w} + [\ell([\underline{u}, \underline{v}]) \cdot \underline{w}]\underline{N}$$

Now, if we use formula (1.80): $[\underline{u}, \underline{v}](\underline{w}) = \underline{u}\,\underline{v}(\underline{w}) - \underline{v}\,\underline{u}(\underline{w})$, we get from the expressions above

$$R(\underline{u}, \underline{v})\underline{w} - [b(\underline{v}, \underline{w})\ell(\underline{u}) - b(\underline{u}, \underline{w})\ell(\underline{v})] + \{[D_{\underline{u}}(\ell(\underline{v})) - D_{\underline{v}}(\ell(\underline{u})) - \ell([\underline{u}, \underline{v}])] \cdot \underline{w}\}\underline{N} = \underline{0}$$

[We used also the Ricci lemma for $\underline{u}(\ell(\underline{v}) \cdot \underline{w})$ and $\underline{v}(\ell(\underline{u}) \cdot \underline{w})$.]

Since in the relation above the tangent and the normal components must vanish separately, we obtain exactly (3.66) and (3.67). Q.E.D.

DEFINITION 3.9 The relations (3.66) and (3.67) are called the integrability conditions. Namely, (3.66) is called the Gauss condition and (3.67) is the Peterson-Mainardi-Codazzi (PMC) condition.

It is clear that (3.66) and (3.67) hold iff they hold for the vectors $\underline{x}_\alpha$ of the basis of TS, because of the linearity properties of the respective expressions. Moreover, since the left-hand side of (3.67) is antisymmetric with respect to $\underline{u}$ and $\underline{v}$, it suffices to express (3.67) for $\underline{u} = \underline{x}_1$, $\underline{v} = \underline{x}_2$ only, and this will contain the whole PMC condition. Since we shall have to set the coefficients of $\underline{x}_1$ and $\underline{x}_2$ in the following expression equal to zero, we finally obtain two relations, which are the following:

$$\text{(PMC)} \quad b^\lambda_{1,2} - b^\lambda_{2,1} = b^\sigma_2 \Gamma^\sigma_{1\sigma} - b^\sigma_1 \Gamma^\lambda_{\sigma 2} \qquad \lambda, \sigma = 1, 2$$

To discuss the Gauss condition (3.66), let us first scalar-multiply it by an arbitrary field $\underline{t}$, which yields the equivalent condition

$$R(\underline{t}, \underline{w}, \underline{v}, \underline{u}) = b(\underline{u}, \underline{t})b(\underline{v}, \underline{w}) - b(\underline{u}, \underline{w})b(\underline{v}, \underline{t}) \tag{3.68}$$

where we have put

$$R(\underline{t}, \underline{w}, \underline{v}, \underline{u}) = [R(\underline{u}, \underline{v})\underline{w}] \cdot \underline{t} \tag{3.69}$$

a relation that defines the covariant curvature tensor of S of the type (0,4).

By (3.64), the components of the tensor defined by (3.69) are

$$R_{\delta\gamma\beta\alpha} = g_{\delta\lambda} R^\lambda{}_{\gamma\beta\alpha}$$

which is just the operation of lowering the first index. By (3.65) this implies that

$$R(\underline{t}, \underline{w}, \underline{v}, \underline{u}) = R(\underline{v}, \underline{u}, \underline{t}, \underline{w})$$

$$R(\underline{t}, \underline{w}, \underline{v}, \underline{u}) = -R(\underline{w}, \underline{t}, \underline{v}, \underline{u})$$

$$R(\underline{t}, \underline{w}, \underline{v}, \underline{u}) = -R(\underline{t}, \underline{w}, \underline{u}, \underline{v})$$

and the only independent condition provided by (3.68) is

$$R(\underline{x}_1, \underline{x}_2, \underline{x}_2, \underline{x}_1) = b_{11}b_{22} - b_{12}^2 \tag{3.70}$$

or, by using the total curvature of S,

$$K = \frac{R(\underline{x}_1, \underline{x}_2, \underline{x}_2, \underline{x}_1)}{\det g} = -\frac{R_{1212}}{\det g} \tag{3.71}$$

Hence this is the only independent Gauss integrability condition, from which we also have

THEOREM 3.3 The total curvature of a surface depends only on the first fundamental form of the surface.

This is a remarkable fact, and Theorem 3.3 is known in differential geometry as the Theorema Egregium of Gauss.

Now we are able to formulate the following basic theorem.

THEOREM 3.4 (Bonnet Fundamental Theorem for Surfaces in E^3) Let $U = \{(u^1, u^2)\}$ be a connected and simply connected domain in $\mathbb{R}^2$, and assume that $g_{\alpha\beta}(u^\gamma)$, $b_{\alpha\beta}(u^\gamma)$ are differentiable functions on U, which satisfy: (1) $g_{\alpha\beta} = g_{\beta\alpha}$, and $g_{\alpha\beta}\xi^\alpha\xi^\beta \geq 0$ with equality holding iff $\xi^\alpha = 0$; (2) $b_{\alpha\beta} = b_{\beta\alpha}$; and (3) the Gauss and PMC conditions. Then there is an immersion $f: U \to \mathbb{R}^3$, which defines a surface in E^3 having the coefficients of the metric tensor $g_{\alpha\beta}$ and the coefficients of the second fundamental form $b_{\alpha\beta}$ (with respect to the local parameterization defined by f). This surface is unique up to a motion in space.

Proof (Sketch): If we take in (3.57) $\underline{v} = \underline{x}_\beta$, $\underline{w} = \underline{x}_\alpha$, and if we add the Weingarten equations (3.20), we see that the Darboux frame $\underline{x}_1$, $\underline{x}_2$, $\underline{N}$ of the parameterization of a surface necessarily satisfies the following system of partial differential equations (the Gauss-Weingarten equations):

$$\begin{aligned} \underline{x}_{\alpha\beta} &= \Gamma^\lambda_{\alpha\beta}\underline{x}_\lambda + b_{\alpha\beta}\underline{N} \\ \underline{N}_\alpha &= -b^\beta_\alpha \underline{x}_\beta \end{aligned} \tag{3.72}$$

Under the hypotheses of the theorem, we can compute the coefficients appearing in (3.72), and then consider (3.72) as a system of equations with the unknowns $\underline{x}_1$, $\underline{x}_2$, $\underline{N}$.

In the theory of such systems, one shows that the first thing to do is to verify that the equations of the system imply the equality of the higher-order derivatives: $\underline{x}_{\alpha\beta\gamma} = \underline{x}_{\alpha\gamma\beta}$, $\underline{N}_{\alpha\beta} = \underline{N}_{\beta\alpha}$. It is obvious that the Gauss and PMC conditions assure that (3.72) implies $\underline{x}_{\alpha\beta\gamma} = \underline{x}_{\alpha\gamma\beta}$. In fact, it suffices to use the computation that led to the Gauss and PMC conditions for the vectors

$\underline{x}_\alpha$, $\underline{x}_\beta$. Moreover, the reader can prove as an exercise, by differentiating the Weingarten equations, that $\underline{N}_{\alpha\beta} = \underline{N}_{\beta\alpha}$ are also implied by the Gauss and the PMC conditions.

Under these circumstances, there is an existence and uniqueness theorem to the effect that every set of initial data $\underline{x}^0_\alpha$, $\underline{N}^0$ associated with a point $(u_0^\alpha) \in U$ uniquely defines a solution $\underline{x}_\alpha$, $\underline{N}$ of (3.72) over U. In addition, if the initial data satisfy the conditions

$$\underline{x}^0_\alpha \cdot \underline{x}^0_\beta = g_{\alpha\beta}(u^\gamma_0) \quad \underline{N}^0 \cdot \underline{x}^0_\alpha = 0 \quad (\underline{N}^0)^2 = 1 \quad (\underline{x}^0_1, \underline{x}^0_2, \underline{N}^0) > 0 \tag{3.73}$$

then similar conditions will hold for the solution $\underline{x}_\alpha$, $\underline{N}$ at every point of U. This can be proven by defining for this solution the associated functions

$$\xi_{\alpha\beta} = \underline{x}_\alpha \cdot \underline{x}_\beta - g_{\alpha\beta} \qquad \eta_\alpha = \underline{N} \cdot \underline{x}_\alpha \qquad \lambda = \underline{N}^2$$

and by showing that (3.72) implies

$$d\xi_{\alpha\beta} = 0 \qquad d\eta_\alpha = 0 \qquad d\lambda = 0$$

Then, because of (3.73), $\xi_{\alpha\beta} = 0$, $\eta_\alpha = 0$, and $\lambda = 1$. The condition $(\underline{x}_1, \underline{x}_2, \underline{N}) > 0$ follows by continuity from the last condition (3.73). The computation is once again left for the reader as an exercise.

Finally, let us define

$$\underline{x}(u) = \int_{u_0}^{u} \underline{x}_\alpha \, du^\alpha + \underline{x}_0 \tag{3.74}$$

where $\underline{x}_\alpha$ is the foregoing solution of (3.72), $u \in U$, $u_0 \in U$ is the initial point, and the integral is a curvilinear integral taken on an arbitrary path that joins u_0 to u in U. (This is well defined since U is simply connected, and the integral does not depend on the path.)

It is clear that (3.74) defines the function f and the surface whose existence was stated by the fundamental theorem. Any other surface that satisfies the conditions of the theorem must also be a solution of (3.72) and (3.74), but possibly corresponds to different initial data. Then the two systems of initial data differ from each other by a motion in space (the proof is left to the reader). Therefore, the two surfaces differ as well from each other by that motion. Q.E.D.

Let us also prove the following "global" uniqueness result:

THEOREM 3.5 Let S and S' be two connected oriented surfaces embedded in E^3. Then an orientation-preserving diffeomorphism $\varphi: S \to S'$ such that, for corresponding points and vectors, the fundamental forms of the two surfaces take the same values, respectively, exists if and only if S is sent onto S' by some motion in E^3.

Proof: The sufficiency part is obvious: If the motion exists, its restriction to S will be φ.

To prove the necessity part, we first note that since the existing φ is an orientation-preserving diffeomorphism, we can find parameterizations about every pair of corresponding points such that the Darboux frames of the two surfaces will correspond to each other by φ.

Now put $S = \cup_{\alpha \in A} V_\alpha$, $S' = \cup_{\alpha \in A} V'_\alpha$, where A is some set of indices, and V_α, V'_α correspond by φ and have corresponding parameterizations with connected and simply connected domain. Let us also assume that every pair V_{α_1}, V_{α_2} and $V_{\alpha'_1}$, $V_{\alpha'_2}$ has a connected intersection. Then, by Theorem 3.4, there is for each α a motion M_α in E^3 such that $M_\alpha(V_\alpha) = V'_\alpha$. Moreover, if $V_\alpha \cap V_\beta \neq \emptyset$, we must have $M_\alpha = M_\beta$. (Why?) Since S is connected, this implies that all the motions M_α coincide with a fixed motion M, and $S' = M(S)$. Q.E.D.

Theorems 3.4 and 3.5 prove the very important fact that the euclidean geometric properties of a surface (i.e., those which are invariant under motions in space), are potentially contained in the two fundamental tensors g and b. Therefore, these tensors constitute a complete system of local invariants of the surface. They are not independent, and are related by the integrability conditions of Gauss and PMC.

EXERCISES

3.43 Compute the Christoffel symbols $\Gamma^\lambda_{\alpha\beta}$ of a surface with respect to an explicit parameterization.

3.44 Find the relation between the Christoffel symbols of the second kind with respect wot two parameterizations whose ranges have a nonempty intersection.

3.45 Let $f: S \to \mathbb{R}$ be a differentiable function defined on the surface S. Consider the vector field $\underline{\xi}$ whose local components are $\xi^\alpha = g^{\alpha\beta} f_\beta$ ($f_\beta = \partial f/\partial u^\beta$). Compute the covariant derivatives of $\underline{\xi}$ with respect to the basic vector field $\underline{x}_\gamma$ ($\gamma = 1, 2$).

3.46 Prove that if S is a plane, $R_{\alpha\beta\gamma\delta} = 0$.

3.47 Compute the Christoffel symbols $\Gamma^\lambda_{\alpha\beta}$ and the components $R_{\alpha\beta\gamma\delta}$ of a surface S with respect to an isothermic parameterization [defined in Exercise 3.15 as a parameterization such that $ds^2 = \lambda(u, v)(du^2 + dv^2)$].

3.48 Let $\underline{v}$ be a tangent vector field and ω be a tangent covector field of a surface S. Set

$$(D_{\underline{v}}\,\omega)(\underline{u}) = \underline{v}(\omega(\underline{u})) - \omega(D_{\underline{v}}\underline{u})$$

where $\underline{u}$ is an arbitrary tangent vector field on S. Prove that $D_{\underline{v}}\omega$

defines a new tangent covector field on S. (This is called the covariant derivative of ω.) Compute $D_{\underline{v}}\omega$ by means of the local components ω_α and v^α, and the Christoffel symbols.

3.49 Let $\underline{v}$ be a tangent vector field on a surface S, and let t be a tensor field of type (0, q). Define

$$(D_{\underline{v}}t)(\underline{u}_1, \ldots, \underline{u}_q) = \underline{v}(t(\underline{u}^1, \ldots, \underline{u}^q)) - \sum_{i=1}^{q} t(\underline{u}^1, \ldots, \underline{u}^{i-1}, D_{\underline{v}}\underline{u}^i, \underline{u}^{i+1}, \ldots, \underline{u}^q)$$

Prove that $D_{\underline{v}}t$ defines a new tensor field of type (0, q) on S. (This is the covariant derivative of t.) Provide a generalization of this procedure for tensor fields of type (p, q). Prove that $D_{\underline{v}}g = 0$ for any $\underline{v}$. (Of course, g is the first fundamental tensor of S.)

3.50 Prove that the curvature tensor of a circular cylinder vanishes.

3.51 Show that if one uses a parameterization with orthogonal parametric lines, the Theorema Egregium yields the following formula for the total curvature of the surface:

$$K = -\frac{1}{2\sqrt{EG}}\left[\left(\frac{E_v}{\sqrt{EG}}\right)_v + \left(\frac{G_u}{\sqrt{EG}}\right)_u\right] \qquad (F = 0)$$

3.52 If a parameterization $\underline{x}(u^\alpha)$ ($\alpha = 1, 2$) of a surface is such that $g_{11} = g_{22} = 1$, $g_{12} = \cos\theta$, the parametric net is called a Tchebyshef net. In such a case, show that the Theorema Egregium yields $K = -\theta_{12}/\sin\theta$. ($\theta_{12} = \partial^2\theta/\partial u^1\,\partial u^2$.)

3.53 Prove that the PMC integrability conditions of a surface S can be put in the form

$$2L_v = \frac{EN + GL}{EG}E_v \qquad 2N_u = \frac{EN + GL}{EG}G_u$$

if the parameters u, v are such that the lines of curvature are parametric lines (F = 0, M = 0). (L, M, N are the components of the second fundamental tensor.)

3.54 Write the Gauss and PMC conditions explicitly for a surface with respect to an isothermic parameterization as considered in Exercise 3.47.

3.55 Prove that there are surfaces such that

$$g_{11} = g_{22} = e^{u^1} \qquad g_{12} = 0 \qquad b_{12} = 0$$

and their total curvature vanishes.

3.56 Prove that no surface $\underline{x} = \underline{x}(u^\alpha)$ ($\alpha = 1, 2$) can have:

(a) $g_{11} = g_{22} = 1,\ g_{12} = 0,\ b_{11} = 1,\ b_{22} = -1,\ b_{12} = 0$

(b) $g_{11} = 1,\ g_{12} = 0,\ g_{22} = \cos^2 u,\ b_{11} = \cos^2 u,\ b_{12} = 0,\ b_{22} = 1$

3.57 Prove that the sphere is the only connected surface such that its first and second fundamental tensors are, respectively, the second and first fundamental tensors of another surface. [Hint: Use the PMC conditions for the two surfaces with respect to parameterizations with the lines of curvature as parametric lines (Exercise 3.53). Show that every point of the surface must be an umbilic, since otherwise $K = 0$, which is impossible here.]

3.4 CURVES ON SURFACES: GEODESIC LINES

In this section we study the euclidean geometric invariants of a curve C defined by a path of S. Let us consider a point $p \in C$. We can define, as for every curve, the natural parameter s, which is the arc length, and get thereby a parameterization with an invariant parameter. Next we shall associate with C at p some orthonormal frame and use it as a moving frame. That is, its derivatives will provide us with the basic invariants.

The frame to be considered is obtained as follows. First, we take the unit tangent vector of C at p, which we shall denote here by $\underline{t}$ ($\underline{t} = \underline{a}$ of Sec. 2.2); then we consider at p the unit vector $\underline{m} = \underline{N} \times \underline{t}$, which is the unit normal vector of C in the tangent plane of S; finally, we consider the unit normal vector $\underline{N}$ of S at p. $(\underline{t}, \underline{m}, \underline{N})$ is a positively oriented orthonormal frame, and we call it the Darboux-Ribaucour frame of $C \subset S$ at p. [As usual, we are concerned with an oriented S, and $\underline{N}$ is already chosen. In particular, over the range of a parameterization about p, $\underline{N}$ is given by (3.11).]

Since the Darboux-Ribaucour frame is orthonormal, the derivatives of its vectors have, by Lemma 2.1, expressions of the form

$$\frac{d\underline{t}}{ds} = \kappa_g \underline{m} + \kappa_n \underline{N} \qquad \frac{d\underline{m}}{ds} = -\kappa_g \underline{t} + \tau_g \underline{N} \qquad \frac{d\underline{N}}{ds} = -\kappa_n \underline{t} - \tau_g \underline{m} \tag{3.75}$$

(By Lemma 2.1 the coefficients form an antisymmetric 3×3 matrix.) The formulas (3.75) are called the Darboux-Ribaucour formulas, and the coefficients are designated as follows: κ_n, the normal curvature; τ_g, the geodesic torsion; and κ_g, the geodesic curvature of $C \subset S$ at p.

Because of the geometric character of $\underline{t}$, $\underline{m}$, $\underline{N}$, ds, the coefficients κ_n, τ_g, and κ_g are invariants as well. Moreover, it is easy to see, by comparing with the fundamental theorem for curves (Theorem 2.2), that κ_n, τ_g, κ_g form a complete system of invariants, which define (locally) $C \subset S$ up to a motion in space. We shall not emphasize this result, which is quite similar to Theorem 2.2, but we shall develop some more facts related to the invariants themselves.

Let us begin by considering the existing relations between the frame $(\underline{t}, \underline{m}, \underline{N})$ and the Frenet frame $(\underline{a}, \underline{b}, \underline{c})$ of C at p. If we denote by ϵ the oriented angle between $\underline{b}$ and $\underline{N}$ (orientation in E^3, i.e., more exactly, in the normal plane of C with respect to $\underline{a}$), these relations are given by (see Fig. 3.5)

$$\underline{t} = \underline{a} \quad \underline{m} = \underline{b} \sin\epsilon - \underline{c} \cos\epsilon \quad \underline{N} = \underline{b}\cos\epsilon + \underline{c}\sin\epsilon \tag{3.76}$$

$$\underline{a} = \underline{t} \quad \underline{b} = \underline{m}\sin\epsilon + \underline{N}\cos\epsilon \quad \underline{c} = -\underline{m}\cos\epsilon + \underline{N}\sin\epsilon \tag{3.77}$$

Now, it is easy to prove

PROPOSITION 3.12 If κ and τ are the ordinary curvature and torsion of C at p, the following relations hold:

$$\kappa_n = \kappa\cos\epsilon \quad \tau_g = \tau + \frac{d\epsilon}{ds} \quad \kappa_g = \kappa\cos\epsilon \tag{3.78}$$

Proof: By differentiating the first relation (3.76) with respect to s, we obtain

$$\kappa\underline{b} = \kappa_g\underline{m} + \kappa_n\underline{N}$$

If in this relation we replace $\underline{b}$ according to the second and third formulas of (3.76), we get the first and third relations of (3.78).

Similarly, by differentiating the second relation (3.76) and using the Frenet and Darboux-Ribaucour formulas, we obtain the second relation of (3.78). (The details are left for the reader.)

REMARK The second formula of (3.78) is meaningful at noninflectional points of C, since otherwise τ is not defined.

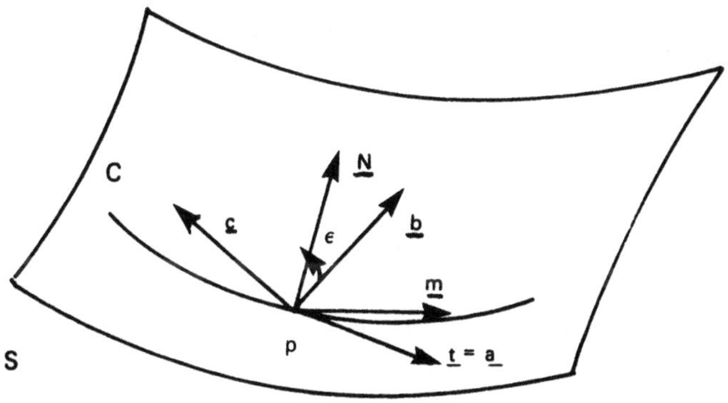

Figure 3.5

On the other hand, we have the following expected result:

PROPOSITION 3.13 The invariants of a curve on a surface can be expressed by means of the fundamental forms of the surface.

Proof: By multiplying the first formula (3.75) by $\underline{N}$ (scalar multiplication), and by replacing $\underline{t} = \underline{a} = d\underline{x}/ds$, where $\underline{x}$ is the radius vector expressed by a parameterization of S and the derivative is along C, we get

$$\kappa_n = \frac{\underline{\underline{N}} \cdot d^2\underline{\underline{x}}}{ds^2} = \frac{\Psi}{\Phi} = \frac{b_{\alpha\beta}\, du^\alpha\, du^\beta}{g_{\lambda\mu}\, du^\lambda\, du^\mu} = \frac{b(\dot{\underline{\gamma}}, \dot{\underline{\gamma}})}{g(\dot{\underline{\gamma}}, \dot{\underline{\gamma}})} = b(\dot{\underline{\gamma}}, \dot{\underline{\gamma}}) \tag{3.79}$$

where $\underline{\gamma}: (a, b) \to S$ is the path defining C, and $\dot{\underline{\gamma}} = d\underline{\gamma}/ds$ is the derivative of the mapping $\underline{\gamma}$. More exactly, $\dot{\underline{\gamma}}$ is the unit tangent vector of C viewed in T_pS, and we used the fact that $\dot{\underline{\gamma}} = \underline{x}_\alpha(du^\alpha/ds)$. Of course, since this is a unit vector, we have $g(\dot{\underline{\gamma}}, \dot{\underline{\gamma}}) = 1$.

Then, by multiplying the last formula in (3.75) by $\underline{m}$, we get

$$\begin{aligned} \tau_g &= -\underline{m} \cdot \frac{d\underline{\underline{N}}}{ds} = -\left(\underline{N}, \underline{t}, \frac{d\underline{\underline{N}}}{ds}\right) = \left(\underline{N}, \frac{d\underline{\underline{N}}}{ds}, \frac{d\underline{\underline{x}}}{ds}\right) \\ &= \frac{c(\dot{\underline{\gamma}}, \dot{\underline{\gamma}})}{\sqrt{\det g}} = \frac{J(g, b)(\dot{\underline{\gamma}}, \dot{\underline{\gamma}})}{\sqrt{\det g}} = \frac{J(\Phi, \Psi)}{\sqrt{\det g} \cdot \Phi} \end{aligned} \tag{3.80}$$

In this computation, we used $\underline{m} = \underline{N} \times \underline{t}$, $c = J(g, b)$ is the tensor defined by (3.50), and the last expression follows by considering $\dot{\underline{\gamma}} = \underline{x}_\alpha(du^\alpha/ds)$; $J(\Phi, \Psi)$ is one-fourth of the jacobian of the two fundamental forms Φ and Ψ with respect to du^1, du^2.

Finally, by multiplying the first formula of (3.75) by $\underline{m}$, we get

$$\kappa_g = \underline{m} \cdot \frac{d\underline{\underline{t}}}{ds} = \left(\underline{N}, \underline{t}, \frac{d\underline{\underline{t}}}{ds}\right)$$

Because of the factor $\underline{N}$, we may replace $d\underline{t}/ds$ here by its tangent component Tg $(d\underline{t}/ds)$ without affecting the result, and by using the tensor ϵ defined by (3.9), we obtain

$$\kappa_g = \epsilon\left(\underline{t}, \operatorname{Tg}\left(\frac{d\underline{\underline{t}}}{ds}\right)\right) \tag{3.81}$$

But

$$\frac{d\underline{\underline{t}}}{ds} = \frac{d}{ds}\left(\underline{x}_\lambda \frac{du^\lambda}{ds}\right) = \underline{x}_\lambda \frac{d^2u^\lambda}{ds^2} + \underline{x}_{\lambda\mu} \frac{du^\lambda}{ds} \frac{du^\mu}{ds}$$

and by using the Gauss equation for $\underline{x}_{\lambda\mu}$, we get

$$\operatorname{Tg} \frac{d\underline{\underline{t}}}{ds} = \left(\frac{d^2u^\beta}{ds^2} + \Gamma^\beta_{\lambda\mu} \frac{du^\lambda}{ds} \frac{du^\mu}{ds}\right) \underline{x}_\beta$$

Hence from (3.81) we obtain

$$\kappa_g = \epsilon_{\alpha\beta} \frac{du^\alpha}{ds}\left(\frac{d^2u^\beta}{ds^2} + \Gamma^\beta_{\lambda\mu} \frac{du^\lambda}{ds}\frac{du^\mu}{ds}\right) \tag{3.82}$$

and by replacing the components $\epsilon_{\alpha\beta}$ by (3.12),

$$\kappa_g = \frac{\sqrt{\det g}}{\Phi^{3/2}} \begin{vmatrix} du^1 & du^2 \\ d^2u^1 + A & d^2u^2 + B \end{vmatrix} \tag{3.83}$$

where

$$A = \Gamma^1_{\lambda\mu} du^\lambda du^\mu \qquad B = \Gamma^2_{\lambda\mu} du^\lambda du^\mu$$

This ends the proof of Proposition 3.13, and the desired expressions are (3.79), (3.80), and (3.83).

COROLLARY 3.1 The geodesic curvature κ_g depends only on the first fundamental form of the surface.

Now we would like to discuss each one of the invariants considered. We begin with κ_n. It follows from (3.79) that $\kappa_n = 0$ iff C is an asymptotic line of S, and the first formula of (3.78) again shows that in this case the osculating plane of C is the tangent plane of S, since $\epsilon = \pi/2$. The first formula of (3.78) also shows that if a surface S contains a segment of a straight line, this must be an asymptotic line, since $\kappa = 0$ implies that $\kappa_n = 0$.

Another consequence of (3.79) is that κ_n depends only on the tangent direction of C at p, which leads to the following considerations. Let us take the intersection of a neighborhood of $p \in S$ with the plane defined by $\underline{N}$ and $\underline{t}$. It is geometrically clear (and we suggest that the reader prove this analytically by using an explicit parameterization of S at p) that this intersection is a curve $\overline{C}$ with the same tangent direction as C. $\overline{C}$ is called the <u>normal section</u> of S corresponding to C. Hence $\kappa_n(\overline{C}) = \kappa_n(C)$. But $\underline{b}(\overline{C}) = \pm\underline{N}$, and $\epsilon(\overline{C})$ is either 0 or π, which implies that $\kappa_n(\overline{C}) = \pm\kappa(\overline{C})$. It follows that

$$|\kappa_n(C)| = \kappa(\overline{C}) \qquad (\geq 0!) \tag{3.84}$$

which confers a geometric meaning on the normal curvature.

Let us define the <u>center of normal curvature</u> K of C to be the center of curvature of $\overline{C}$. Considering separately the cases $\epsilon(\overline{C}) = 0$ and $\epsilon(\overline{C}) = \pi$, we see that K lies on the normal line of S at p, and it has the radius vector $\underline{x} + (1/\kappa_n)\underline{N}$. With the notation $1/\kappa_n = \rho_n$, $1/\kappa = \rho$, the first formula (3.78) gives $\rho = \rho_n \cos\epsilon$. This proves

THEOREM 3.6 (Meusnier) The center of curvature of C is the orthogonal projection of its normal center of curvature onto the principal normal of C.

From the theory of curves, we know that $\bar{C}$ and $\underline{b}(\bar{C})$ are on the same side of the tangent of $\bar{C}$. Once again, by taking separately $\epsilon(\bar{C}) = 0$ and $\epsilon(\bar{C}) = \pi$, we see that $\bar{C}$, hence S, and $\underline{N}$ are on the same side of the tangent plane of S for those directions for which $\kappa_n > 0$, and S and $\underline{N}$ are on different sides of the tangent plane if $\kappa_n < 0$. This allows a study of the geometric shape of the surface about p, by studying the variation of κ_n. First we prove

THEOREM 3.7 (Euler) Let p be a nonumbilical point of S, and let $\tilde{\kappa}_1$, $\tilde{\kappa}_2$ be the normal curvatures of the principal directions at p. Then the normal curvature of the direction making an angle θ with the direction of normal curvature $\tilde{\kappa}_1$ is given by

$$\kappa_n = \tilde{\kappa}_1 \cos^2 \theta + \tilde{\kappa}_2 \sin^2 \theta \tag{3.85}$$

Proof: Since p is not an umbilic, we can choose a parameterization at p whose parametric lines are the lines of curvature of S (we are using Proposition 3.1), and we shall have $g_{12} = 0$, $b_{12} = 0$. Then, if $\tilde{\kappa}_1$ is the normal curvature of $u^1 = \text{const}$ and $\tilde{\kappa}_2$ is the normal curvature of $u^2 = \text{const}$, we get

$$\kappa_n = \frac{b_{11}(du^1)^2 + b_{22}(du^2)^2}{g_{11}(du^1)^2 + g_{22}(du^2)^2} \qquad \tilde{\kappa}_1 = \frac{b_{22}}{g_{22}} \qquad \tilde{\kappa}_2 = \frac{b_{11}}{g_{11}} \tag{3.86}$$

and, using (3.6) and (3.13),

$$\cos\theta = \frac{(g_{22})^{\frac{1}{2}}\, du^2}{[g_{11}(du^1)^2 + g_{22}(du^2)^2]^{\frac{1}{2}}} \qquad \sin\theta = \frac{(g_{11})^{\frac{1}{2}}\, du^1}{[g_{11}(du^1)^2 + g_{22}(du^2)^2]^{\frac{1}{2}}} \tag{3.87}$$

Now we establish (3.85) by simply checking it with these values. Q.E.D.

Moreover, using the same parameterization, we see that $\tilde{\kappa}_1$, $\tilde{\kappa}_2$ of (3.86) are exactly the solutions of Eq. (3.43); that is, they are actually the principal curvatures κ_1, κ_2 of S at p. Hence the Euler formula (3.85) becomes

$$\kappa_n = \kappa_1 \cos^2 \theta + \kappa_2 \sin^2 \theta \tag{3.88}$$

Note that this formula is trivially verified at an umbilic.

The Euler formula (3.88) provides us with the geometric meaning of κ_1 and κ_2 as the normal curvatures of the lines of curvature and as the extremal values of the normal curvature κ_n at point p. In fact, if we assume that $\kappa_1 \le \kappa_2$, (3.88) implies that $\kappa_1 \le \kappa_n \le \kappa_2$.

Finally, this allows us to discuss the signs of $\kappa_n(\theta)$, hence the geometric shape of S at p, and we find again the three known cases of elliptic ($\kappa_1\kappa_2 > 0$), hyperbolic ($\kappa_1\kappa_2 < 0$), and parabolic ($\kappa_1\kappa_2 = 0$) points. (We leave it to the reader to work out the details of this discussion.)

We shall not dwell on the geodesic torsion τ_g. Let us only note, by (3.80), that τ_g is also a function of the tangent direction of C alone (at a fixed p), and that $\tau_g = 0$ characterizes the lines of curvature of S.

On the other hand, we shall embark now on a rather lengthy discussion of the geodesic curvature κ_g and the equation $\kappa_g = 0$. Since (3.82) and (3.83) contain the second-order derivatives, κ_g of two different tangent curves at p may be different. κ_g is a very important invariant because of Corollary 3.1, and because it generalizes the curvature k of a plane curve as defined in Sec. 2.5. Indeed, if S is a plane and if u^1, u^2 are cartesian coordinates, we have $g_{\alpha\beta} = \delta_{\alpha\beta}$ and $\Gamma^{\alpha}_{\beta\gamma} = 0$. Then formula (3.83) reduces to (2.88).

Before starting our discussion of κ_g, we would like to clarify the significance of Tg $(d\underline{t}/ds)$ encountered in formula (3.81). This is especially worthwhile since it is related to important geometric notions. Let us consider a curve C defined by a path $\gamma: (a,b) \to S$ with the parameter $\tau \in (a,b)$. Let us also consider a vector field $\underline{v}: (a,b) \to E^3$ which is tangent to S along C in the sense that $\underline{v}(\tau) \in T_{\gamma(\tau)}S$ for every $\tau \in (a,b)$. Then $d\underline{v}/d\tau$ is again a vector in E^3, and we should like to compute its projection onto $T_{\gamma(\tau)}S$, that is, <u>its tangent component</u>.

By $\underline{v} = v^{\alpha}\underline{x}_{\alpha}$, and using the Gauss equations of Sec. 3.3, we obtain

$$\frac{d\underline{v}}{d\tau} = \frac{D\underline{v}}{d\tau} + b_{\alpha\beta} v^{\alpha} \frac{du^{\beta}}{d\tau} \underline{N}$$

where we denote

$$\frac{D\underline{v}}{d\tau} = \left(\frac{dv^{\alpha}}{d\tau} + \Gamma^{\alpha}_{\lambda\mu} v^{\lambda} \frac{du^{\mu}}{d\tau}\right)\underline{x}_{\alpha} \tag{3.89}$$

and this is the tangent component we want.

The vector (3.89) is called the <u>covariant derivative of</u> v <u>along</u> γ (or C), and if we consider that $\underline{v}$ is the restriction of C of a field defined on an open neighborhood of C, we see by (3.61) that

$$\frac{D\underline{v}}{d\tau} = \frac{v^{\alpha}}{\beta(du^{\beta}/d\tau)\,\underline{x}_{\alpha}} \tag{3.90}$$

DEFINITION 3.10 $\underline{v}$ is called a <u>parallel vector field</u> along C if $D\underline{v}/d\tau = 0$.

This notion was introduced by T. Levi-Civita, and it is called the <u>Levi-Civita parallelism</u>. It is a natural geometric notion in the following sense: Suppose that the surface is the universe of some two-dimensional beings who are unable to realize the third dimension; for them $d\underline{v}/d\tau$ reduces to $D\underline{v}/d\tau$, and parallelism (i.e., the constancy of the vector $\underline{v}$) is defined by setting this derivative equal to zero.

It follows from (3.89) that $\underline{v}$ is parallel iff

$$\frac{dv^{\alpha}}{d\tau} + \Gamma^{\alpha}_{\lambda\mu} v^{\lambda} \frac{du^{\mu}}{d\tau} = 0 \tag{3.91}$$

which is a system of first-order differential equations. Hence every vector

$\underline{v}_0 \in T_{\gamma(\tau_0)}S$ can be considered as the initial data of (3.91) and it defines a unique parallel field along γ. This operation is known as the Levi-Civita parallel translation of $\underline{v}_0$ along γ. Finally, we note the obvious fact that the Levi-Civita parallelism depends only on the first fundamental form of S.

We therefore have now the geometric meaning of Tg ($d\underline{t}/ds$):

$$\mathrm{Tg}\,\frac{d\underline{t}}{ds} = \frac{D\underline{t}}{ds}$$

and (3.81) becomes

$$\kappa_g = \epsilon\left(\underline{t}, \frac{D\underline{t}}{ds}\right) \tag{3.92}$$

Moreover, it is natural to consider such curves C on S for which $D\underline{t}/ds = \underline{0}$, that is, curves whose unit tangent field is a parallel field. Indeed, in the plane, these are just the straight lines, so the curves considered should play the role of the "straight lines" of a general surface.

DEFINITION 3.11 A curve C on S whose unit tangent field is a parallel field is called a geodesic line of S.

PROPOSITION 3.14 C is a geodesic line iff $\kappa_g(C) = 0$.

Proof: Formula (3.92) shows that the geodesic curvature of a geodesic line vanishes. Conversely, since ϵ is antisymmetric, $\kappa_g = 0$ implies by (3.92) that

$$\frac{D\underline{t}}{ds} = \rho\underline{t}$$

for some multiplier ρ. Now take $\underline{t}^2 = 1$ and differentiate it with respect to s. This gives

$$0 = \underline{t}\,\frac{d\underline{t}}{ds} = \underline{t}\,\frac{D\underline{t}}{ds} = \rho\underline{t}^2 = \rho$$

and we obtain $D\underline{t}/ds = \underline{0}$. Q.E.D.

COROLLARY 3.2 A line C of S is a geodesic line iff the principal normal of C equals the normal of S at every point of C. (Or, equivalently, the osculating plane of C contains the normal of S.)

Proof: This follows from Proposition 3.14 and formulas (3.78), which show that $\kappa_g = 0$ iff $\epsilon = 0$ or π. Q.E.D.

In particular, every straight line segment that belongs to S is a geodesic line of S, since it has an indeterminate osculating plane. (Or, since $\kappa_g = \kappa \sin \epsilon$ and $\kappa = 0$.) The sphere offers an interesting example; namely, for every great circle of it both the principal normal and the normal of the sphere are exactly the line of the radius at the respective point. Hence every great circle is a geodesic of the sphere.

PROPOSITION 3.15 The geodesic lines are characterized by

$$\frac{d^2u^\alpha}{ds^2} + \Gamma^\alpha_{\sigma\tau}\frac{du^\sigma}{ds}\frac{du^\tau}{ds} = 0 \tag{3.93}$$

They depend on the first fundamental form only, and for every $p_0 \in S$ and every $\underline{v}_0 \in T_{p_0}S$ there is one and only one geodesic arc passing through p_0 and tangent to the direction of $\underline{v}_0$ at that point.

Proof: Equation (3.93) follows from (3.91) by taking $\tau = s$, $v^\alpha = du^\alpha/ds$, and this shows also that the geodesic lines depend only on the metric of S. The last assertion is nothing other than the determination of a solution of (3.93) by initial data. Note that (3.93) is a system of equations of the second order, and that "arc" has the meaning of Sec. 2.1.

To apply this, let us prove that a sphere has no other geodesic lines except the arcs of the great circles. Indeed, every point and tangent direction determine a great circle: the intersection of the sphere with the plane passing through the center of the sphere and through the given direction. Hence every geodesic must be an arc of a great circle.

Now we should like to consider another aspect of the geodesic lines.

DEFINITION 3.12 A local parameterization $\underline{x}$ of S is called semigeodesic if the first fundamental tensor satisfies the conditions $g_{11} = 1$, $g_{12} = g_{21} = 0$, with respect to that parameterization.

PROPOSITION 3.16 For every point $p_0 \in S$ and for every geodesic arc C through p_0, there is a semigeodesic parameterization of S at p_0 such that C belongs to the parametric lines $u^2 = \text{const}$.

Proof: (See Fig. 3.6.) Let Γ be an arc of S passing through p_0 and orthogonal to C at this point. (For example, Γ can be the geodesic defined by p_0 and by the direction orthogonal to C and tangent to S.) Then let us consider the family $\tilde{C}$ of the geodesic arcs which are defined by the points of Γ and the tangent directions orthogonal to Γ. (These exist by Proposition 3.15.)

Now we consider the tangent vector field of $\tilde{C}$, and its orthogonal vector field, and by Proposition 3.1 we construct a parameterization $\underline{x}$ whose parametric lines are tangent to these two vector fields. It is obvious that this is correctly defined in some neighborhood of p_0, and moreover, we can choose the parameters such that $\tilde{C}$ is $u^2 = \text{const}$. Note that C is an element of $\tilde{C}$.

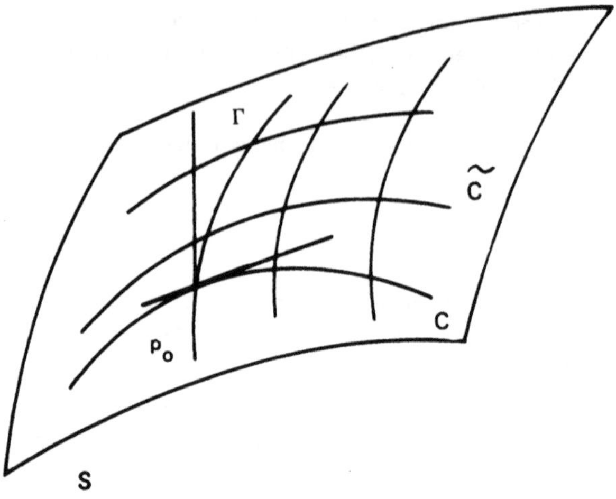

Figure 3.6

Since the parametric lines are orthogonal, we have $g_{12} = 0$. Since u^2 = const satisfies the equations (3.93), we must have $\Gamma_{11}^2 = 0$, that is, by (3.63), $\partial g_{11}/\partial u^2 = 0$ and g_{11} is a function of u^1 only. By setting

$$\tilde{u}^1 = \int \sqrt{g_{11}}\, du^1 \qquad \tilde{u}^2 = u^2$$

we get a new parameterization at p_0 which is obviously semigeodesic and satisfies the conclusions of Proposition 3.16. Q.E.D.

REMARK If Γ is chosen to be a geodesic and defined by $u^1 = 1$, it satisfies (3.93), which yields the <u>initial data</u> $(\partial g_{22}/\partial u^1)\big|_{u^1=0} = 0$. We may take u^2 as the natural parameter of Γ, which gives $g_{22}(0, u^2) = 1$. (The reader is asked to prove the details.)

This preparation enables us to prove another important property of geodesic lines:

PROPOSITION 3.17 For every two sufficiently nearby points $p, q \in S$ which are joined by a geodesic line C, C is the shortest path between p and q.

Proof: Let us take the semigeodesic parameterization of Proposition 3.16 at p, and assume that "sufficiently nearby" means that q belongs to the range of this parameterization. Then if $u^\alpha(\tau)$ is some path $\tilde{C}$ joining p and q, its length is given by

$$\ell(\tilde{C}) = \int_p^q [(u^1)'^2 + g_{22}(u^2)'^2]^{\frac{1}{2}} \, d\tau \geq \int_p^q du^1 = u^1(q) - u^1(p)$$

On the other hand,

$$\ell(C) = \int_p^q du^1 = u^1(q) - u^1(p)$$

Hence $\ell(\tilde{C}) \geq \ell(C)$. Q.E.D.

However, we should note that we did not discuss the existence of a geodesic line between two points of a surface. One can prove that two sufficiently nearby points p, $q \in S$ are always joined by a unique geodesic ℓ of minimal length. This situation is preserved if q runs along the previous geodesic line ℓ until some position m, which is called the first conjugate point of p along ℓ. Afterward, the geodesic is no longer minimal. For example, if S is a sphere, the conjugate of a point along a great circle is the diametrically opposite point.

One can also prove that if a surface is such that the equations (3.93) have global solutions [i.e., solutions for $s \in (-\infty, +\infty)$], then every two points are joined by a minimal geodesic line. The surfaces whose geodesics are defined by $s \in (-\infty, +\infty)$ are called complete. For example, every compact surface is complete.

The geodesic curvature is important for arbitrary curves on a surface as well. Let us prove one interesting result.

PROPOSITION 3.18 Let C be a curve on S, $\underline{t}$ its unit tangent vector field, $\underline{m}$ the field $\underline{N} \times \underline{t}$, and s the natural parameter. Then we have

$$\frac{D\underline{t}}{ds} = \kappa_g \underline{m} \qquad \frac{D\underline{m}}{ds} = -\kappa_g \underline{t} \tag{3.94}$$

Proof: It follows from the Darboux-Ribaucour formulas (3.75) that $\kappa_g \underline{m}$ is the tangent component of $d\underline{t}/ds$, and $-\kappa_g \underline{t}$ is the tangent component of $d\underline{m}/ds$. Q.E.D.

These formulas are a generalization of the Frenet formulas for plane curves, and we call them the intrinsic Frenet formulas for curves on S.

REMARK We end this section with some additional remarks. On several occasions we noted that various elements depend only on the first fundamental form of the surface. This is important because such elements are preserved by diffeomorphisms of the surface that preserve the first fundamental form. Such diffeomorphisms are called isometries. It follows that the elements which depend only on the first fundamental form are rather closely related to the surface: not only motions, but arbitrary isometries preserve them. This part of the geometry of a surface, defined by its first fundamental form, is called the intrinsic geometry of the surface. The properties that depend also on the second fundamental form constitute the extrinsic geometry of the surface.

EXERCISES

3.58 Compute the normal curvature of the parametric lines of the torus with the parameterization of Exercise 3.23. Discuss the classification of the points of the torus and the corresponding geometric form of the surface.

3.59 Compute the vectors $\underline{t}$, $\underline{m}$, $\underline{N}$ and the invariants κ_n and τ_g of a loxodrome of a sphere. (See Exercise 3.11 for the definition of a loxodrome.)

3.60 Let p be a point of a sphere S with center O. Denote by γ the circle of intersection of S with the orthogonal bisector plane of the segment Op. Compute the vectors $\underline{t}$, $\underline{m}$, $\underline{N}$ and the invariants κ_n, τ_g, κ_g of γ. Is γ a geodesic line of S?

3.61 Let p be a point of a surface S. Prove that:

(a) The sum of the normal curvatures for any pair of orthogonal directions at p equals the mean curvature of S at p.

(b) The mean curvature of S at p is also given by the formula $H = (1/\pi)\int_0^\pi \kappa_n(\theta)\, d\theta$, where θ is the angle of the corresponding direction with a fixed tangent direction at p.

3.62 Let d be the direction that makes an angle θ with the principal direction of normal curvature κ_1. Prove the following Bonnet formula for the geodesic torsion of d:

$$\tau_g = \frac{1}{2}(\kappa_1 - \kappa_2)\sin 2\theta$$

3.63 Let γ be an asymptotic line without inflection points of a surface S. Prove that one has (the Beltrami-Enneper theorem)

$$|\tau| = \sqrt{-K}$$

at every hyperbolic point, where τ is the torsion of γ and K is the total curvature of S. [Hint: Use the result of Exercise 3.62.]

3.64 For a point p of a surface S, consider the locus of the points q in the tangent plane $\pi_p S$ which have the polar coordinates $(\sqrt{|1/\kappa_n|}, \theta)$ with respect to the pole p and the axis defined by the direction of normal curvature κ_1. This locus is called the Dupin indicatrix of S at p.

Discuss the nature of the Dupin indicatrix at an elliptic, hyperbolic, and parabolic point of S. Find an interpretation of the asymptotic directions and of the conjugate direction in relation to the Dupin indicatrix. [Hint: Take cartesian coordinates in $\pi_p S$, and show, using the Euler formula, that the indicatrix is composed of conics. The asymptotic and conjugate directions are then the usual ones for conics.]

3.65 (Theorem of Joachimstal) Assume that the oriented surfaces S_1 and S_2 intersect along a connected regular curve C with no inflection points. Prove that any two of the following properties imply the third property:

1. C is a line of curvature on S_1.
2. C is a line of curvature on S_2.
3. The angle between the tangent planes of S_1 and S_2 is constant along C.

[Hint: Use the fact that $\tau_g = 0$ for the lines of curvature, and also use formulas (3.78).]

3.66 Assume that surfaces S_1 and S_2 are tangent to each other along a regular arc C. Prove that the Levi-Civita parallel translation of vectors along C is the same for the two surfaces.

3.67 Prove that the Levi-Civita parallel translation preserves the length of a vector and the angle between two vectors. [Hint: Show that scalar products are preserved. This is a consequence of the Ricci lemma for covariant derivatives, studied in Sec. 3.3.]

3.68 Consider on a sphere S a curvilinear triangle abc, where a is the pole, bc is the arc of the corresponding equator, and ab and ac are arcs of great circles (meridians). Prove that if we translate in parallel (in the sense of Levi-Civita) the unit tangent vector of ac at a along ac, then along cb, and finally, along ba, we arrive at the unit vector tangent to ba at a. [Hint: Use the fact that the great circles are geodesics, and refer to Exercise 3.66.]

3.69 Prove that if the oriented surfaces S_1 and S_2 are tangent to each other along a regular arc C, then C has the same geodesic curvature on the two surfaces. In particular, C is a geodesic line of S_1 iff C is a geodesic line of S_2.

3.70 Let C be a regular arc with no inflection point on a surface S. Prove that any two of the following properties imply the third property:

1. C is a line of curvature.
2. C is situated in a plane.
3. The angle between the principal normal of C and the normal of S is constant.

3.71 Prove that a curve C is both an asymptotic line and a geodesic line on a surface S iff C is a segment of a straight line.

3.72 Prove that the parametric lines of the surface

$$x^1 = \frac{a}{2}(u^1 + u^2) \quad x^2 = \frac{b}{2}(u^1 - u^2) \quad x^3 = \frac{u^1u^2}{2}$$

are geodesic lines.

3.73 Prove that every curve is a geodesic line on the surface generated by its binormals (which is a regular surface in some neighborhood of the curve).

3.74 Assume that a surface S has a parameterization $\underline{x} = \underline{x}(u, v)$ with respect to which $g_{12} = 0$ and $g_{11} = g_{22} = U(u) + V(v)$. (This is called a Liouville surface.) Prove that the geodesic lines of S are defined by the equation

$$\int \frac{du}{\sqrt{U - c}} = \pm \int \frac{dv}{\sqrt{V + c}} + c'$$

where c and c' are real constants.

3.75 Let $\underline{x} = \underline{x}(u, v)$ be an isothermal parameterization of a surface S (Exercise 3.15). Prove that if one family of parametric lines have constant geodesic curvature, the same is true for the other family of parametric lines. Write the intrinsic Frenet formulas for a curve u = const.

3.76 Let $\varphi: S \to S'$ be a differentiable mapping of surfaces, whose differentials preserve the angle between two arbitrary tangent vectors. Then φ is called a conformal transformation. Prove that:

(a) A conformal transformation cannot have singular points.

(b) φ is a conformal transformation iff for any parameterizations $\underline{x}: U \to S$ of S and $\varphi \circ \underline{x}: U \to S'$ of S' $(U \subset \mathbb{R}^2)$, one has $g'_{\alpha\beta} = \rho g_{\alpha\beta}$, $\rho > 0$ (g is the first tensor of S and g' is the first tensor of S'); every isothermal parameterization (Exercise 3.15) establishes a conformal transformation between its range and a plane domain.

3.77 Let $\varphi: S \to S'$ be a differentiable mapping of surfaces whose differentials preserve the length of every tangent vector. Then φ is called a local isometry or, if φ is also a diffeomorphism, an isometry between S and S'. Prove that:

(a) φ is a conformal transformation and has no singular points.

(b) φ is a local isometry iff, for any parameterizations $\underline{x}: U \to S$, $\varphi \circ \underline{x}: U \to S'$ $(U \subset \mathbb{R}^2)$, one has $g'_{\alpha\beta} = g_{\alpha\beta}$ (g, g' are the first fundamental tensors of S, S', respectively).

(c) The local intrinsic geometry of S and S' is the same, and in particular, S and S' have the same total curvature at corresponding points.

Give an example of a differentiable mapping $\varphi: S \to S'$ which is not an isometry, yet is such that the total curvature of S and S' is the same at corresponding points. [Hint: One example for the last part of the exercise is provided by the correspondence between points with the same values of the parameters of the surfaces defined by

(S) $x^1 = u \cos v,\ x^2 = u \sin v,\ x^3 = \ln u,$

(S') $x^1 = u \cos v,\ x^2 = u \sin v,\ x^3 = v.$]

3.5 SOME PARTICULAR CLASSES OF SURFACES

In previous sections we examined surfaces from a general viewpoint. Here we consider some particular classes of surfaces.

Surfaces generated cinematically

Intuitively, we can think of a surface as the locus described by a curve moving in space. We would like to consider two such motions. First, we consider a curve in E^3 that is turning around some axis. The locus so obtained is called a surface of revolution, and we want to see how this definition fits in with our usual notion of a surface.

It is clear that the sections of a surface of revolution cut by planes perpendicular to the axis are circles, and these will be called parallel lines. The surface is cut by planes through the axis forming lines which are called meridians. It is clear that a surface of revolution can always be considered to be generated by the rotation of a meridian.

Then let us choose the frame in E^3 such that Ox^3 is the axis of rotation and the initial meridian is a curve in x^1Ox^3. Moreover, assume that this curve is $x^1 = u^1$, $x^2 = 0$, $x^3 = \varphi(u^1)$. Then we see from Fig. 3.7 that the coordinates of a point M of the surface are given by

$$x^1 = u^1 \cos u^2 \quad x^2 = u^1 \sin u^2 \quad x = \varphi(u^1) \quad u^1 \geq 0 \tag{3.95}$$

Since these look like a parameterization that is regular for $u^1 \neq 0$, we can now provide a definition that is rigorous from the viewpoint of the previous sections.

DEFINITION 3.13 A surface S in E^3 is called a surface of revolution if there is a regular parameterization of the form (3.95) at every point $p \in S$.

We do not ask for only one global parameterization (3.95). However, the transition relations between two such parameterizations are of special form. Indeed, if

$$x^1 = \tilde{u}^1 \cos \tilde{u}^2 \quad x^2 = \tilde{u}^1 \sin \tilde{u}^2 \quad x^3 = \psi(\tilde{u}^1) \tag{3.96}$$

is another such parameterization (regularity implies that $u^1 > 0$, $\tilde{u}^1 > 0$),

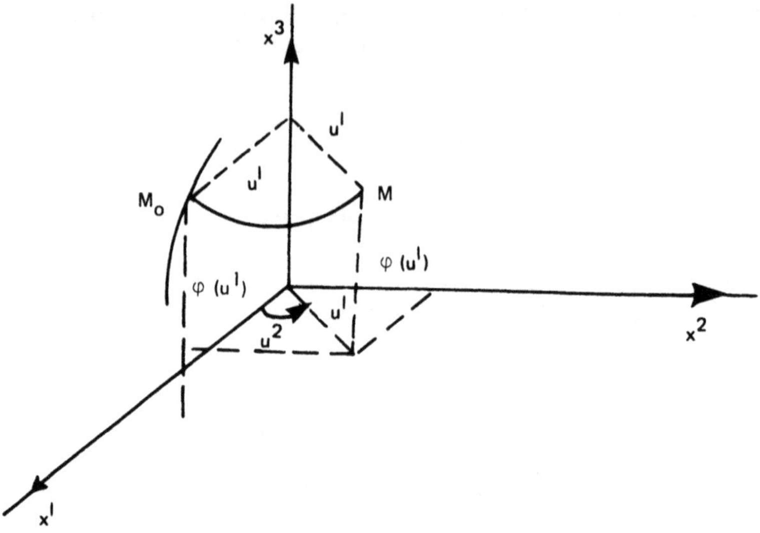

Figure 3.7

then, on every connected component of the intersection of their ranges, we have by a simple comparison

$$\tilde{u}^1 = u^1 \qquad \tilde{u}^2 = u^2 + 2h\pi \qquad \psi(u^1) = \varphi(u^1) \tag{3.97}$$

where h is an integer that may vary with the connected component above. A surface of revolution cannot be covered by only one range of a parameterization (3.95) since for a regular parameterization u^2 must be restricted to an open set of values.

REMARK Equations (3.97) show that a surface of revolution is orientable.

Another reason for the appearance of more than one parameterization (3.95) for a surface of revolution can be the fact that the meridian which generates the surface may have no global explicit equation. (However, it always has local explicit equations.) For example, if the generating meridian has global parametric equations $x^1 = \psi(u^1)$, $x^2 = 0$, $x^3 = \varphi(u^1)$, we get similarly (Fig. 3.7) the following type of parameterizations of a surface of revolution:

$$x^1 = \psi(u^1) \cos u^2 \qquad x^2 = \psi(u^1) \sin u^2 \qquad x^3 = \varphi(u^1) \tag{3.98}$$

On the other hand, if the generating meridian has a global implicit equation $F(x^1, x^3) = 0$, $x^2 = 0$, it follows, by computing u^1 in the plane through M and Ox^3 (Fig. 3.7) that the surface has the global implicit equation

$$F([(x^1)^2 + (x^2)^2]^{\frac{1}{2}}, x^3) = 0 \tag{3.99}$$

For instance, if the generating curve is the circle

$$(x^1)^2 + (x^3)^2 - 2ax^1 + b = 0 \quad x^2 = 0$$

with center $(a, 0, 0)$ and radius $\rho = (a^2 - b)^{\frac{1}{2}} < a$, the corresponding surface of revolution is the torus (Fig. 3.8):

$$4a^2[(x^1)^2 + (x^2)^2] = [(x^1)^2 + (x^2)^2 + (x^3)^2 + b]^2$$

For the same torus we also have the following parametric equations of the form (3.98):

$$x^1 = (\rho \cos u^1 + a) \cos u^2 \quad x^2 = (\rho \cos u^1 + a) \sin u^2 \quad x^3 = \rho \sin u^1 \tag{3.100}$$

Now we use the parameterization (3.95) to compute various geometric elements of a surface of revolution. The natural tangent basis is

$$\underline{x}_1(\cos u^2, \sin u^2, \varphi') \quad \underline{x}_2(-u^1 \sin u^2, u^1 \cos u^2, 0) \tag{3.101}$$

where primes denote derivatives with respect to u^1. The unit normal vector is

$$\underline{N} = \frac{\underline{x}_1 \times \underline{x}_2}{|\underline{x}_1 \times \underline{x}_2|} = \left(-\frac{\varphi'}{[1 + (\varphi')^2]^{\frac{1}{2}}} \cos u^2, \ -\frac{\varphi'}{[1 + (\varphi')^2]^{\frac{1}{2}}} \sin u^2, \ \frac{1}{[1 + (\varphi')^2]^{\frac{1}{2}}}\right) \tag{3.102}$$

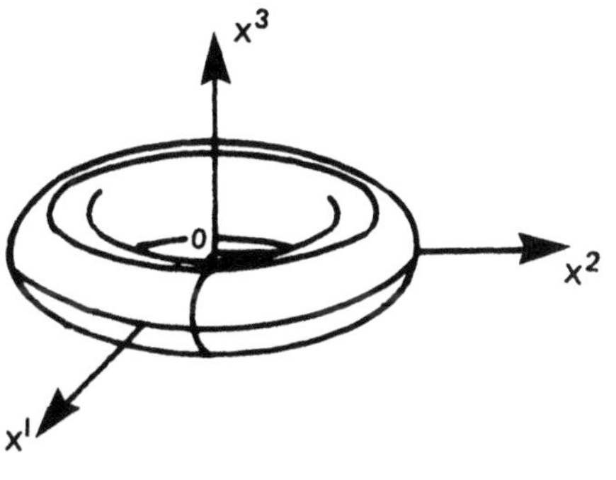

Figure 3.8

The tangent plane has the equation

$$\begin{vmatrix} x^1 - u^1 \cos u^2 & x^2 - u^1 \sin u^2 & x^3 - \varphi(u^1) \\ \cos u^2 & \sin u^2 & \varphi' \\ -u^1 \sin u^2 & u^1 \cos u^2 & 0 \end{vmatrix} = 0$$

The normal line has the equations

$$x^1 = u^1 \cos u^2 + \lambda\varphi' \cos u^2$$

$$x^2 = u^1 \sin u^2 + \lambda\varphi' \sin u^2$$

$$x^3 = \varphi(u^1) - \lambda$$

where λ is a parameter. It is easy to deduce that this line is in the plane defined by M and the Ox^3 axis; hence we have

PROPOSITION 3.19 The normal line of a surface of revolution at a point M is exactly the principal normal of the meridian that passes through M.

The components of the first fundamental form are

$$g_{11} = 1 + \varphi'^2 \quad g_{12} = g_{21} = 0 \quad g_{22} = (u^1)^2 \tag{3.103}$$

The components of the first fundamental form are

$$b_{11} = \frac{\varphi''}{(1 + \varphi'^2)^{\frac{1}{2}}} \quad b_{12} = b_{21} = 0 \quad b_{22} = \frac{u^1\varphi'}{(1 + \varphi'^2)^{\frac{1}{2}}} \tag{3.104}$$

As a consequence of $g_{12} = 0$, $b_{12} = 0$, we have

PROPOSITION 3.20 The lines of curvature of a surface of revolution are exactly the meridians and the parallels.

For the asymptotic lines, we have the differential equation

$$\varphi''(du^1)^2 + u^1\varphi'(du^2)^2 = 0$$

which yields

$$\frac{du^2}{du^1} = \pm\left(\frac{\varphi''}{u^1\varphi'}\right)^{\frac{1}{2}}$$

and therefore

$$u^2 = c \pm \int \left(\frac{\varphi''}{u^1\varphi'}\right)^{\frac{1}{2}} du^1$$

where c is an integration constant. Now the mean curvature and the total curvature of the surface are given by

$$H = \frac{1}{(1 + \varphi'^2)^{\frac{1}{2}}}\left(\frac{\varphi'}{u^1} + \frac{\varphi''}{1 + \varphi'^2}\right) \quad u^1 > 0 \tag{3.105}$$

$$K = \frac{\varphi'\varphi''}{u^1(1 + \varphi'^2)^2} \qquad u^1 > 0 \tag{3.106}$$

This allows us to discuss elliptic hyperbolic and parabolic points on a surface of revolution. For instance, we have

PROPOSITION 3.21 The parallel line generated by an inflection point of the generating meridian consists of parabolic points of the surface.

Proof: This follows from (3.106) since $\varphi'' = 0$ at an inflection point. Q.E.D.

Next, if we are interested in geodesic lines, we should start with the following consequence of Proposition 3.19 and Corollary 3.2:

PROPOSITION 3.22 The meridians of a surface of revolution are geodesic lines.

However, these are not all the geodesics of the surface since we do not have a meridian in every tangent direction. To get all the geodesics we must compute the Christoffel symbols. The results are

$$\begin{aligned} &\Gamma^1_{11} = \frac{\varphi'\varphi''}{1 + \varphi'^2} \quad \Gamma^1_{12} = 0 \quad \Gamma^1_{22} = -\frac{u^1}{1 + \varphi'^2} \\ &\Gamma^2_{11} = 0 \quad \Gamma^2_{12} = \frac{1}{u^1} \quad \Gamma^2_{22} = 0 \end{aligned} \tag{3.107}$$

and hence the differential equations of the geodesics are

$$\begin{aligned} &\frac{d^2u^1}{ds^2} + \frac{\varphi'\varphi''}{1 + \varphi'^2}\left(\frac{du^1}{ds}\right)^2 - \frac{u^1}{1 + \varphi'^2}\left(\frac{du^2}{ds}\right)^2 = 0 \\ &\frac{d^2u^2}{ds^2} + \frac{2}{u^1}\frac{du^1}{ds}\frac{du^2}{ds} = 0 \end{aligned} \tag{3.108}$$

As a consequence, we get

PROPOSITION 3.23 (Clairaut) Let us denote by α the angle between a geodesic line and the parallel line through a point. Then $u^1 \cos\alpha = \text{const}$ along the geodesic line (u^1 is the radius of the parallel).

Proof: We have

$$\cos\alpha = g_{\alpha\beta}\,\xi^\alpha \frac{du^\beta}{ds}$$

where ξ^α are the components of the unit tangent vector of the line $u^1 = \text{const}$, that is, $\xi^1 = 0$, $\xi^2 = 1/\sqrt{g_{22}} = 1/u^1$ $(u^1 > 0)$. That is,

$$u^1 \cos \alpha = (u^1)^2 \frac{du^2}{ds}$$

which, by a differentiation, and in view of the second equation of (3.108), yields $d(u^1 \cos \alpha)/ds = 0$. Q.E.D.

Now let us consider again a generating curve, but a more complicated motion, consisting of a rotation about an axis, and simultaneously, a translation along the same axis by a length that is proportional to the rotation angle. If the generating curve is again $x^1 = u^1$, $x^2 = 0$, $x^3 = \varphi(u^1)$, and the axis is Ox^3, we obviously get a parameterization of the form

$$x^1 = u^1 \cos u^2 \qquad x^2 = u^1 \sin u^2 \qquad x^3 = \varphi(u^1) + hu^2 \tag{3.109}$$

where $h = \text{const}$.

DEFINITION 3.14 A differentiable surface that has a regular parameterization of the form (3.109) at every point is called a helicoid.

The reader is asked to compute the various geometric elements of such surfaces, as was done in the case of a surface of revolution, and to compare the results obtained.

A particularly important case is obtained for $\varphi = k = \text{const}$. The corresponding surface, with parameterizations of the form

$$x^1 = u^1 \cos u^2 \qquad x^2 = u^1 \sin u^2 \qquad x^3 = hu^2 + k \tag{3.110}$$

is called a ruled helicoid, since the lines $u^2 = \text{const}$ are straight lines. This surface has the following interesting property:

PROPOSITION 3.24 The ruled helicoid has vanishing mean curvature.

Proof: From (3.110), we get

$$g_{11} = 1 \qquad g_{12} = 0 \qquad g_{22} = (u^1)^2 + h^2$$

$$b_{11} = 0 \qquad b_{12} = \frac{-h}{[(u^2)^2 + h^2]^{\frac{1}{2}}} \qquad b_{22} = 0$$

whence $H = 0$. Q.E.D.

DEFINITION 3.15 A surface that has vanishing mean curvature is called a minimal surface.

The study of minimal surfaces is beyond the scope of this book, but we should like to emphasize its importance in differential geometry. Minimal surfaces are those which satisfy the necessary condition of an extremum if we are looking for surfaces bounded by a given closed curve and having the smallest area (the Plateau problem).

Ruled and developable surfaces

Classically, a ruled surface is conceived of as a one-parameter family of straight lines, called the generators of the surface. Then if $\underline{y}(\tau)$ is (locally) a curve that intersects the generators, and if $\underline{v}(\tau)$ defines the direction of the generator through $\underline{y}(\tau)$, the equation of the generators is

$$\underline{x} = \underline{y}(\tau) + \lambda\underline{v}(\tau) \tag{3.111}$$

and this can be viewed as the parametric equation of the surface.

Moreover, if we consider two parameterizations (3.111), corresponding to curves $\underline{y}(\tau)$ and $\tilde{\underline{y}}(\tilde{\tau})$, then the intersection of these curves with the generators through the common points p of the ranges of the parameterizations, establishes a correspondence between $\underline{y}(\tau)$ and $\tilde{\underline{y}}(\tilde{\tau})$ (Fig. 3.9). Hence we must have a transition relation of the form $\tilde{\tau} = \varphi(\tau)$.

Therefore, we are led to introduce

DEFINITION 3.16 A surface $S \subset E^3$ is called a ruled surface if it has a regular parameterization of the form

$$\underline{x} = \underline{y}(u^1) + u^2\underline{v}(u^1) \tag{3.112}$$

at every point, such that a transition relation of the form

$$\tilde{u}^1 = \varphi(u^1) \tag{3.113}$$

holds in the common part of the ranges of any two such parameterizations.

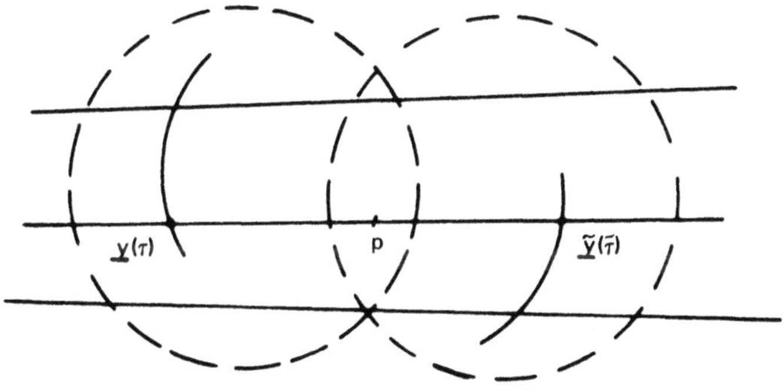

Figure 3.9

DEFINITION 3.17 A maximally connected curve of S whose nonempty intersections with the ranges of the parameterizations (3.112) have equations u^1 = const is called a generator.

This definition is meaningful because of (3.113), and using (3.112) we see that the generators are open segments of straight lines.

PROPOSITION 3.25 The generators of a topologically closed ruled surface $S \subset E^3$ are full straight lines.

Proof: Since S is topologically closed, it cannot contain an open segment that is not a whole line. Q.E.D.

REMARKS

1. Proposition 3.25 implies that (3.112) can be extended to $u^2 \in (-\infty, +\infty)$, that is, to a parameterization whose range contains full straight lines (if S is closed). However, it may happen that this global parameterization has singular points.
2. If we have two parameterizations (3.112) and a common point of their ranges, we must have $\underline{\tilde{v}}(\tilde{u}^1) = \lambda(u^1)\underline{v}(u^1)$ $[\tilde{u}^1 = \varphi(u^1)]$ at that point. This follows from the fact that the generator passing through a given point is uniquely defined.
3. If convenient, we may use in (3.112) unit vectors $\underline{v}$ only. Then we have a transition relation $\underline{\tilde{v}}(\tilde{u}) = \underline{v}(u^1)$.
4. As a matter of fact, a ruled surface is a combined structure consisting of a surface S and a special atlas on S (see Sec. 1.4 for the notion of an atlas). It can happen that a surface admits more than one such atlas (i.e., more than one ruled structure). For example, this happens in the case of the one-sheeted hyperboloid and of the hyperbolic paraboloid.
5. The ruled helicoid (3.110) is an example of a ruled surface.

Now we want to express the main geometric elements of a ruled surface by means of a parameterization (3.112). To simplify the computations, we assume that $|\underline{v}| = 1$ and $|\underline{y}'| = 1$, that is, u^1 is the natural parameter (arc length) of the curve $\underline{y}(u^1)$ ("prime" denotes the derivative with respect to u^1). We also assume that the curve $\underline{y}(u^1)$ is orthogonal to the generators (i.e., $\underline{v} \cdot \underline{y}' = 0$). Clearly, this implies no loss of generality. Then

$$\underline{x}_1 = \underline{y}' + u^2\underline{v}' \qquad \underline{x}_2 = \underline{v} \tag{3.114}$$

$$g_{11} = 1 + 2u^2\underline{y}' \cdot \underline{v}' + (u^2)^2(\underline{v}')^2 \qquad g_{12} = 0 \qquad g_{22} = 1 \tag{3.115}$$

$$\underline{N} = \frac{\underline{y}' \times \underline{v} + u^2\underline{v}' \times \underline{v}}{[1 + 2u^2\underline{y}' \cdot \underline{v}' + (u^2)^2(\underline{v}')^2]^{\frac{1}{2}}} \tag{3.116}$$

$$b_{11} = \frac{(\underline{y}'' + u^2\underline{v}'', \underline{y}' + u^2\underline{v}', \underline{v})}{[1 + 2u^2y'v' + (u^2)^2(v')^2]^{\frac{1}{2}}} \qquad b_{12} = \frac{(\underline{v}', \underline{y}', \underline{v})}{[1 + 2u^2\underline{y}'\underline{v}' + (u^2)^2(\underline{v}')^2]^{\frac{1}{2}}}$$

$$b_{22} = 0 \tag{3.117}$$

Hence, as expected, the generators $u^1 = \text{const}$ are asymptotic lines, and we can say that a surface S is ruled iff one of its families of asymptotic lines consists of straight line segments. As a corollary, we have the following interesting fact:

PROPOSITION 3.26 For a ruled surface, one has $K \leq 0$ everywhere.

Proof: Since the generators offer a real asymptotic direction, we cannot have elliptic points on the surface. Q.E.D.

The following definition introduces an interesting class of surfaces:

DEFINITION 3.18 A surface S is called locally flat if $K = 0$ everywhere.

THEOREM 3.8 A locally flat surface without planar points is ruled, and it has a fixed tangent plane along every generator.

Proof: Take $p \in S$. Since p is not planar, and $K(p) = \kappa_1(p) \cdot \kappa_2(p) = 0$, we must have, for example, $\kappa_1 = 0$, $\kappa_2 \neq 0$. Then p is not umbilical, and there is a parameterization $\underline{x}(u^1, u^2)$ at p whose parametric net is the net of the lines of curvature. Moreover, we may assume that $\kappa_1 = 0$ along the lines $u^1 = \text{const}$ and that u^2 is the arc length along $u^1 = 0$. (If this is not so from the beginning, we replace the parameters by $v^1 = u^1$, $v^2 = \int [g_{22}(0, u^2)]^{\frac{1}{2}} du^2$.)

Using the Weingarten transformation, we obtain

$$\underline{N}_1 = -\kappa_2 \underline{x}_1 \quad \underline{N}_2 = \underline{0}$$

and since $g_{12} = 0$ and $b_{12} = 0$,

$$\underline{x}_1 \cdot \underline{x}_2 = 0 \quad \underline{N}_1 \cdot \underline{x}_2 = 0$$

Since $\underline{N} \cdot \underline{x}_2 = 0$ and $\underline{N}_2 = \underline{0}$, we have by a differentiation $\underline{N} \cdot \underline{x}_{22} = 0$. By differentiating $\underline{N}_1 \cdot \underline{x}_2 = 0$ with respect to u^2 and using $\underline{N}_{22} = \underline{N}_{21} = \underline{0}$ and $\underline{N}_1 = -\kappa_2 x_1$ ($\kappa_2 \neq 0$), we get $\underline{x}_1 \cdot \underline{x}_{22} = 0$.

Finally, note that $\partial(\underline{x}_2^2)/\partial u^1 = 2\underline{x}_2 \cdot \underline{x}_{21} = -2\underline{x}_1 \cdot \underline{x}_{22}$ (by differentiating $\underline{x}_1 \cdot \underline{x}_2 = 0$ with respect to u^2). Hence $\partial(\underline{x}_2^2)/\partial u^1 = 0$, and $\underline{x}_2^2$ does not depend on u^1 (if we consider a connected domain for our parameterization). Since u^2 is the arc length on $u^1 = 0$ we have $\underline{x}_2^2(0, u^2) = 1$ and, therefore, $\underline{x}_2^2(u^1, u^2) = 1$. By a differentiation, this yields $\underline{x}_2 \cdot \underline{x}_{22} = 0$. The three relations $\underline{N} \cdot \underline{x}_{22} = 0$, $\underline{x}_1 \cdot \underline{x}_{22} = 0$, $\underline{x}_2 \cdot \underline{x}_{22} = 0$ imply that $\underline{x}_{22} = 0$, which shows that our parameterization is of the form (3.112).

Furthermore, we must also have the transition relation (3.113) between two such parameterizations since, for both parameterizations, the lines $u^1 = \text{const}$ ($\tilde{u}^1 = \text{const}$) are those defined by $\kappa^1 = \text{const}$ ($\tilde{\kappa}^1 = \text{const}$), and κ^1 is an invariant of the surface ($\kappa_1 = \tilde{\kappa}_1$). This ends the proof of the fact that the surface considered is a ruled surface.

Since, with the notation above, the generators are $u^1 = \text{const}$, and we have $\underline{N}_2 = \underline{0}$, it follows that $\underline{N}$ is constant along a generator, and the tangent plane, which is orthogonal to $\underline{N}$, is also constant along the generator. Q.E.D.

One can see by examples that the results do not hold if S has planar points. On the other hand, one has

THEOREM 3.9 A ruled surface whose tangent plane is constant along the generators is a locally flat surface.

Proof: The hypothesis of the fixed tangent plane yields $\underline{N}_2 = \underline{0}$ (the notation above), and by the Weingarten transformation, $\underline{N}_2 = -\ell \underline{x}_2$. Hence $\underline{x}_2$ is a principal vector with the corresponding principal curvature $\kappa_1 = 0$. Since $K = \kappa_1 \kappa_2$, we get $K = 0$. Q.E.D.

DEFINITION 3.19 A ruled surface with constant tangent plane along the generators is called a developable surface.

Then Theorems 3.9 and 3.8 mean, respectively, that a developable surface is locally flat, and a locally flat surface without planar points is developable.

The local geometric structure of a locally flat surface is determined completely by the following theorem.

THEOREM 3.10 Let $p \in S$ be a point of a locally flat surface. Then we always have one of the following situations:

(a) p belongs to an open neighborhood of S that lies in a plane.

(b) p belongs to an open neighborhood of S that lies on a cylinder.

(c) p belongs to an open neighborhood of S that lies on a cone.

(d) p belongs to an open neighborhood of S that lies on a surface generated by the tangents of some space curve.

(e) p is the limit of a sequence of points that satisfy one of the conditions (b)-(d).

Proof: First, let us consider the case of a planar point p. Then if a whole neighborhood of p consists of planar points, this neighborhood belongs to a plane (Proposition 3.3), and we have situation (a). If p has no such neighborhood, p is the limit of a sequence of nonplanar points (since every neighborhood of p has nonplanar points), and this case will be covered by situation (e).

Now let p be a nonplanar point. By Theorem 3.9 we have, for example, $\kappa_1 = 0$, $\kappa_2 > 0$, which allows us to use a parameterization (3.112) at p whose parametric lines are the lines of curvature, and such that $\underline{v}^2 = 1$, $\underline{y}'^2 = 1$ (u^1 is the arc length on $u^2 = 0$). (Such a parameterization has been considered in the proof of Theorem 3.8.) This implies that b_{12} of (3.117) vanishes and that $\underline{v}$ and $\underline{y}'$ are orthogonal ($g_{12} = 0$). It follows that

$$\underline{v}' = \alpha \underline{y}' + \beta \underline{v}$$

whence, by a scalar multiplication with $\underline{v}$, we deduce that $\beta = 0$. That is,

$$\underline{v}' = \alpha \underline{y}' \qquad \alpha = \underline{y}' \cdot \underline{v}' \tag{3.118}$$

Now, if $\alpha = 0$ on an open neighborhood of p, we have $\underline{v} = \text{const}$, and this neighborhood belongs to a cylinder (which is by definition a ruled surface with parallel generators). This is exactly situation (b). If $\alpha(p) = 0$ and every neighborhood of p has points with a nonvanishing α, p is the limit of a sequence of points with nonvanishing α and we shall again have case (e).

Furthermore, assume that $\alpha(p) \neq 0$ (then this holds on a whole neighborhood of p), and let us look for a curve that is tangent to the straight lines containing the generators of S. Such a curve must have an equation of the form

$$\underline{z} = \underline{y}(u^1) + \lambda(u^1)\underline{v}(u^1) \tag{3.119}$$

with the tangent vector $\underline{y}' + \lambda'\underline{v} + \lambda\underline{v}'$. This is collinear to $\underline{v}$ iff $\lambda = -\alpha$, where α is given by (3.118).

Now we have necessarily the following situations:

1. The point $\underline{z}$ of (3.119) with $\lambda = -\alpha$ is a fixed point (i.e., $\underline{z}' = 0$ on a whole neighborhood of p). Then this neighborhood belongs to a cone (which is defined as a ruled surface whose generator lines pass through a fixed point). This is exactly case (c) of the theorem.
2. $\underline{z}'(p) \neq 0$, which is clearly case (d).
3. $\underline{z}'(p) = 0$ and p is the limit of a sequence of points with $\underline{z}' \neq 0$ in case (e).

This ends the proof of Theorem 3.10. The reader is asked to prove, as an exercise, that a plane, a cylinder, a cone, and a surface generated by the tangents of a space curve are always developable surfaces.

It is more difficult to establish the global geometric structure of the locally flat surfaces. If the surface is required to be complete [i.e., with geodesics defined on $(-\infty, +\infty)$], one can show that it must be either a plane or a cylinder (where the latter may include a plane domain).

Surfaces of constant curvature

We have already encountered the surfaces of vanishing total curvature. Generally, by a surface of constant curvature we understand a surface S with $K = \text{const}$. Here we should like to obtain some information about such surfaces.

Let us take a point $p \in S$ and a semigeodesic parameterization $\underline{x}(u^1, u^2)$ at p (Proposition 3.16). Then

$$g_{11} = 1 \quad g_{12} = 0 \tag{3.120}$$

and we shall also assume, as in the remark following Proposition 3.16, that

$$g_{22}(0, u^2) = 1 \quad \left(\frac{\partial g_{22}}{\partial u^1}\right)_{u^1=0} = 0 \tag{3.121}$$

By means of this parameterization and the Gauss Theorema Egregium [formula (3.71)], we obtain

$$K = -\frac{1}{\sqrt{g_{22}}} \cdot (\sqrt{g_{22}})_{11} \tag{3.122}$$

(The concrete computation is left as an exercise for the reader. One should not forget that the final indices 1 denote derivatives with respect to u^1.)

If $K = \text{const}$, (3.122) yields the differential equation

$$(\sqrt{g_{22}})_{11} + K\sqrt{g_{22}} = 0 \tag{3.123}$$

which can be solved as follows:

1. $K > 0$. Then it is well known that the general solution of (3.123) has the following form:

$$\sqrt{g_{22}} = c_1(u^2)\cos(\sqrt{K}\,u^1) + c_2(u^2)\sin(\sqrt{K}u^1)$$

and if the initial data (3.121) are assumed, we get

$$g_{22} = \cos^2(\sqrt{K}u^1) \tag{3.124}$$

2. $K < 0$. In this case, the solution of (3.123) and (3.121) is

$$g_{22} = \cosh^2(\sqrt{-K}u^1) \tag{3.125}$$

3. $K = 0$. Then the solution of (3.121) and (3.123) is

$$g_{22} = 1 \tag{3.126}$$

This discussion shows that the value of $K = \text{const}$ defines completely the first fundamental form of the surface. Two surfaces of the same constant total curvature have the same first fundamental forms at corresponding points if we consider local correspondences between the two surfaces defined by the equality of the semigeodesic parameters of the surfaces. Because of this fact, we state

THEOREM 3.11 Any two surfaces of the same constant total curvature are locally isometric, and have the same intrinsic geometry.

Theorem 3.11 suggests that we look for some standard models of surfaces of constant curvature. This is very easy for $K = 0$. It suffices to take a plane with the cartesian coordinates (u^1, u^2). Then $g_{11} = g_{22} = 1$, $g_{12} = 0$, and $K = 0$. Consequently, every locally flat surface is locally isometric to a plane. (See other information about this case in our previous discussion of ruled surfaces.)

In the proof of Proposition 3.6 we saw that the principal curvatures at every point of a sphere are both equal to $1/r$, r being the radius of the sphere. Therefore, the total curvature is $1/r^2$, and we can see that if $K > 0$ the sphere of radius $1/\sqrt{K}$ can be taken as the standard model of a surface of constant positive curvature. Consequently, every surface of constant positive curvature is locally isometric to a sphere.

A semigeodesic parameterization of a sphere of radius r is provided by the equations

$$x^1 = r\cos u^1 \cos u^2 \qquad x^2 = r\cos u^1 \sin u^2 \qquad x^3 = r\sin u^1$$

followed by the parameter transformation $\tilde{u}^1 = ru^1$, $\tilde{u}^2 = ru^2$. This suggests that we look for a model of the case $K < 0$ in the class of the surfaces of revolution. We describe the model in a straightforward manner and advise the reader to rederive it by using the general formulas that we gave for surfaces of revolution.

Consider the surface of revolution defined by

$$x^1 = h(u^1)\cos u^2 \qquad x^2 = h(u^1)\sin u^2 \qquad x^3 = k(u^1) \tag{3.127}$$

where

$$h(u^1) = r\exp\left(\frac{u^1}{r}\right) \qquad k(u^1) = \int_0^{u^1}\left[1 - \exp\left(\frac{2u^1}{r}\right)\right]^{\frac{1}{2}} dt \qquad r > 0 \tag{3.128}$$

Then one has

$$g_{11} = 1 \qquad g_{12} = 0 \qquad g_{22} = h^2 \tag{3.129}$$

Hence (3.127) is a semigeodesic parameterization, and formula (3.122) yields $K = -1/r^2$. (See also Exercise 3.34.)

The surface (3.127) and (3.128) is called a pseudosphere, and it follows that every surface of constant negative curvature is locally isometric to a pseudosphere. These surfaces are also important in geometry because they provide us with a model of the Lobatchewsky geometry: If we think of their geodesics as straight lines, we find more than one parallel line through a given point to a given line. [See, for instance, Jacobs (1974).]

Global results can be obtained from supplementary hypothesis. For instance, we have

PROPOSITION 3.27 Let S be a compact surface in E^3. Then there is at least one point $p \in S$ with $K(p) > 0$.

Proof: (Sketch): Since S is compact it is bounded, and we define the infimum r of the radii of the closed disks centered at the origin and containing S. Let Σ_r be the sphere of radius r centered at the origin. Then it is obvious that $\Sigma_r \cap S \neq \emptyset$; let us consider a point $p \in \Sigma_r \cap S$. We also see from the definition of r that Σ_r and S have the same unit normal vector $\underline{N}$ at p [otherwise, S gets out from the disk bounded by Σ_r (see Fig. 3.10) and, therefore, the same tangent plane].

Then let us consider a common tangent vector $\underline{v}$ of S and Σ_r, and the plane α defined by $(\underline{v}, \underline{N})$. α cuts S and Σ_r by curves whose normal curvatures have the same sign. [Indeed, S and Σ_r are on the same side of the tangent plane (Fig. 3.10).] Moreover, since S is interior to Σ_r the section

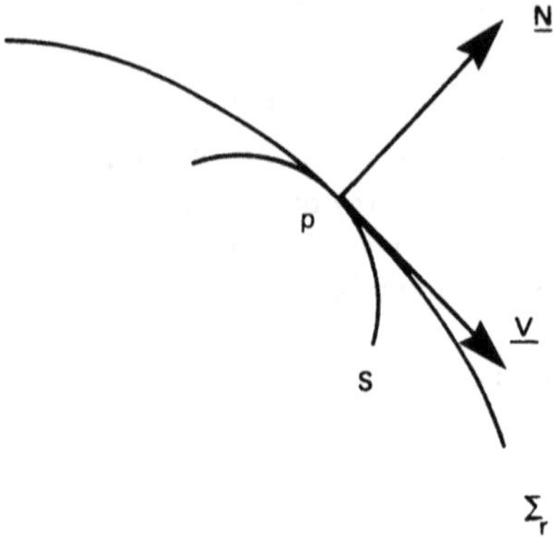

Figure 3.10

$S \cap \alpha$ has a "larger curvature" (i.e., it has $|\kappa_n| \geq 1/r$, where $1/r$ is the normal curvature of any curve on Σ_r). If we apply this to the principal vectors $\underline{v}$ of S at p, we get $K(p) = \kappa_1(p)\kappa_2(p) \geq 1/r^2 > 0$. Q.E.D.

COROLLARY 3.3 There is no compact surface in E^3 with strictly negative, and in particular, with constant negative sectional curvature.

It is possible to prove, moreover, that there are no complete surfaces in E^3 with constant negative sectional curvature. It is also possible to prove that every complete connected simply connected surface in E^3 such that $K \leq 0$ (not necessarily constant) is diffeomorphic to the whole plane $\mathbb{R}^2$.

Other results: Every connected simply connected complete locally flat surface is globally isometric to a plane, and every connected simply connected complete surface of constant positive sectional curvature is globally isometric to a sphere.

EXERCISES

3.78 Write the equations of a surface of revolution if the meridian is a straight line, a circle, and an ellipse.

3.79 Find the curves lying on a surface of revolution and cutting the meridians at a constant angle. (These are called the loxodromes of the surface.) Compute the normal curvature of these curves.

3.80 Find the curves of a surface of revolution that are conjugate to the loxodromes cutting the meridians at the angle α. (See the definition of loxodromes in Exercise 3.79.)

3.81 Compute the geodesic curvature of the parallels of a surface of revolution. Discuss the possibility of the parallels being geodesic lines.

3.82 Determine (locally) the surfaces of revolution that have vanishing mean curvature (i.e., are minimal surfaces).

3.83 Determine the locally flat surfaces of revolution.

3.84 Determine (locally) the surfaces of revolution that have constant negative sectional curvature.

3.85 Determine (locally) the surfaces of revolution that have constant positive sectional curvature.

3.86 Write the equations of a helicoid generated by a straight line, a circle, and an ellipse.

3.87 The intersection curves of a helicoid with planes passing through the axis are called meridians. Prove that the meridians of a helicoid $\mathscr{H}$ are geodesics iff $\mathscr{H}$ is either a ruled helicoid or a surface of revolution.

3.88 Each point of the generating curve of a helicoid describes a helix. Prove that:
(a) The total curvature of a helicoid is constant along a helix.
(b) A curve that cuts the helices at a right angle is a geodesic line.

3.89 Prove that every helicoid is locally isometric with a surface of revolution.

3.90 Prove that if a connected surface S has more than two ruled structures, S lies on a plane.

3.91 Prove that if one family of asymptotic lines of a surface consists of segments of straight lines, the surface is ruled.

3.92 Assume that the ruled surface $\mathscr{R}$ has a parameterization $\underline{x} = \underline{y}(s) + \lambda\underline{v}(s)$, where s is the arc length of $\underline{y}(s)$, $|\underline{v}| = 1$, and $\underline{v}'(s)$ never vanishes. Prove that there is a unique (not necessarily regular) curve γ on $\mathscr{R}$ such that the vectors $\underline{v}$ are parallel along γ in the sense of Levi-Civita. (γ is called the line of striction of $\mathscr{R}$, and its intersection point with every generator is called the central point of the generator.) [Hint: Require that the tangent part of $\underline{v}'$ be zero.]

3.93 Consider the surface defined by the parameterization

$$x^1 = \frac{1}{4}(u^3 - 3uv^2 - 3u) \qquad x^2 = \frac{1}{4}(2u^2v - v^3 + 3v) \qquad z = \frac{3}{4}(v^2 - u^2)$$

where u, v are parameters (the Enneper surface). Prove, using Exercise 3.91, that this is a ruled surface, and find its line of striction (Exercise 3.92).

3.94 Consider a ruled surface $\mathcal{R}$ defined by $\underline{x} = \underline{y}(\tau) + \lambda \underline{v}(\tau)$. Prove that $\mathcal{R}$ is a developable surface iff $(\underline{y}', \underline{v}, \underline{v}') = 0$. [Hint: Use the fact that the tangent plane is constant along a generator.]

3.95 Consider the ruled surfaces defined by the principal normals and by the binormals of a space curve. What are the conditions for these surfaces to be developable?

3.96 Find the line of striction of a developable surface.

3.97 Prove that a curve C of a surface S is a line of curvature iff the normals of S along C form a developable surface. [Hint: Use the Darboux-Ribaucour formulas for C.]

3.6 THE GAUSS-BONNET FORMULA. COMPACT SURFACES

The global differential geometry of surfaces is a highly interesting subject. In order to give some of its flavor, we shall briefly discuss one famous topic: the Gauss-Bonnet theorem and its relations with the classification of compact surfaces. This classification is much more complicated than the classification of curves discussed in Sec. 2.4. We follow Klingenberg's exposition [Klingenberg (1977)].

Let us consider the range of a semigeodesic parameterization $\underline{x}(u^1, u^2)$ of a surface S, which means that we have

$$g_{11} = 1 \quad g_{12} = 0 \quad g_{22} = G \tag{3.130}$$

and correspondingly,

$$\Gamma^1_{11} = \Gamma^2_{11} = \Gamma^1_{12} = 0 \quad \Gamma^2_{11} = \frac{1}{\sqrt{G}}\frac{\partial\sqrt{G}}{\partial u^1} \quad \Gamma^1_{22} = -\frac{1}{2}\frac{\partial G}{\partial u^1}$$
$$\Gamma^2_{22} = \frac{1}{\sqrt{G}}\frac{\partial\sqrt{G}}{\partial u^1} \tag{3.131}$$

and

$$K = -\frac{1}{\sqrt{G}}\frac{\partial^2\sqrt{G}}{(\partial u^1)^2} \tag{3.132}$$

As usual, the natural basis of the tangent plane is $\underline{x}_1$ and $\underline{x}_2$, where by (3.130), $|\underline{x}_1| = 1$, $|\underline{x}_2| = \sqrt{G}$.

Now consider a curve C on S contained in the range of $\underline{x}$ and either defined on [a,b] or on $\mathbb{R}$, and periodic (i.e., closed). We denote by $\underline{t}$ the unit tangent field of C. Then there is a decomposition

$$\underline{t} = \underline{x}_1 \cos\theta(s) + \frac{\underline{x}_2}{\sqrt{G}} \sin\theta(s) \tag{3.133}$$

where $s \in [a,b]$ ($s \in \mathbb{R}$) is the natural parameter of C and θ is the oriented angle from $\underline{x}_1$ to $\underline{t}$. Moreover, following exactly the same reasoning as in Proposition 2.5, we see that $\theta(\tau)$ can be considered as a differentiable function on $[a,b]$ (or $\mathbb{R}$), and it is defined up to the addition of a term $2h\pi$, where h is a constant integer. On the other hand, we have as usual

$$\frac{d\underline{t}}{ds} = \underline{x}_1 \frac{du^1}{ds} + \underline{x}_2 \frac{du^2}{ds} \tag{3.134}$$

and, by comparing with (3.133),

$$\frac{du^1}{ds} = \cos\theta(s) \qquad \frac{du^2}{ds} = \frac{1}{\sqrt{G}} \sin\theta(s) \tag{3.135}$$

The second derivatives are obtainable from this by a new differentiation:

$$\begin{aligned} \frac{d^2u^1}{ds^2} &= -\sin\theta \cdot \dot{\theta} \\ \frac{d^2u^2}{ds^2} &= \frac{1}{\sqrt{G}} \cos\theta \cdot \dot{\theta} - \frac{(\sqrt{G})_1}{G} \sin\theta\cos\theta - \frac{(\sqrt{G})_2}{G^{3/2}} \sin^2\theta \end{aligned} \tag{3.136}$$

where the dot denotes a derivative with respect to s and the indices 1 and 2 in the second relation denote derivatives with respect to u^1 and u^2, respectively. Now if we replace in formula (3.82) the expressions given by (3.130), (3.131), (3.135), and (3.136) and perform the computation, we get for the geodesic curvature of C the expression

$$\kappa_g = \frac{d\theta}{ds} + \frac{\partial\sqrt{G}}{\partial u^1} \frac{du^2}{ds} \tag{3.137}$$

Now, we proceed with the preparation for the main result. Let us consider a polygonal domain A in the plane $\mathbb{R}^2 = \{(u^1, u^2)\}$. Then it is meaningful to speak of a diffeomorphism $F: A \to S$ if we consider it as the restriction of a diffeomorphism of an open neighborhood of A to a corresponding domain in S. (In fact, this is a diffeomorphism of manifolds similar to the one in Sec. 1.6.) The image of such a diffeomorphism F is called a polygonal domain P of S.

If such a domain is given, the restrictions of F to the edges of A define curves in S, and the curves obtained form a polygon in S, called the boundary of the domain and denoted by ∂P. A and its edges will be oriented so as to allow the utilization of Green's formula, which is well known from calculus. By means of F, we then get an orientation of P and ∂P. The images of the vertices of A are called the vertices of P, and the tangents at the two edges that meet at a vertex form an external angle α (see Fig. 3.11). We assume that the reader is familiar with integrals on surfaces, and we denote by $d\sigma$ the area of an element of S.

Now, we can prove

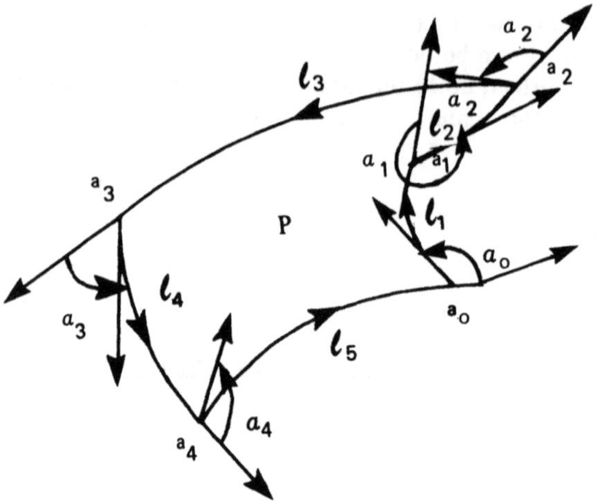

Figure 3.11

THEOREM 3.12 For every polygonal domain P of S, the following formula holds:

$$\iint_P K\,d\sigma = 2\pi - \sum_j \alpha_j - \int_{\partial P} \kappa_g(s)\,ds \tag{3.138}$$

This is called the Gauss-Bonnet formula.

Proof:

(i) In the first step, we assume that the domain P is in the range of a semigeodesic parameterization, in which (3.132) holds. Then, using successively the definition of the surface integral, the Green formula, the relation (3.137), and the definition of the curvilinear integral, we obtain

$$\begin{aligned}\iint_P K\,d\sigma &= \iint_A K\sqrt{\Delta}\,du^1\,du^2 = -\iint_A \frac{\partial^2\sqrt{G}}{(\partial u^1)^2}\,du^1\,du^2 \\ &= -\int_{\partial A} \frac{\partial\sqrt{G}}{\partial u^1}\,du^2 = \int_{\partial P}(d\theta - \kappa_g\,ds) = \int_{\partial P} d\theta - \int_{\partial P}\kappa_g\,ds\end{aligned} \tag{3.139}$$

Here $\int_{\partial P} d\theta$ is understood as follows: ∂P consists of the differentiable curves $\ell_1, \ldots, \ell_n$, where each ℓ_i ($i = 1, \ldots, n$) corresponds to an edge of A and joins the vertices a_{i-1}, a_i ($a_n = a_0$) (see Fig. 3.11). Then for each ℓ_i we have a differentiable function $\theta^i(s)$ defined by (3.133) (the addition of a constant $2h\pi$ does not change the result). Then the needed integral is

$$\int_{\partial P} d\theta = \sum_{i=1}^{n} \int_{\ell_i} d\theta^i = \sum_{i=1}^{n} [\theta^i(a_i) - \theta^i(a_{i-1})]$$

If we add to the integral above the sum $\Sigma_{i=1}^{n} \alpha_i$ of the external angles of the polygon ($\alpha_n = \alpha_0$), we get the complete variation of the angle θ of the tangent $\underline{t}$ with the direction $\underline{x}_1$, while going around the boundary of P, with the following convention: All the $\theta^i(s)$ are chosen differentiable and are fixed by the initial values $\theta^1(a_0)$ arbitrary, $\theta^{i+1}(a_i) = \theta^i(a_i) + \alpha_i$.

If S is a plane, this variation is 2π by the theorem of turning tangents (Theorem 2.6) for a piecewise differentiable simple closed curve with positive orientation. In the case of a general surface S, this variation must also have the form $2h\pi$ for some integer h, since after a turn we arrive at the same geometric tangent, and by (3.135) we must have the same $\cos\theta$, $\sin\theta$.

Now, if we define

$$\Phi_\tau = (du^1)^2 + [\tau + (1-\tau)G](du^2)^2 \qquad 0 \leq \tau \leq 1$$

we get a continuous deformation of the metric of S to the metric of a plane. Correspondingly, we get for the variation of θ (which can be computed by means of the metric only) the value $2h(\tau)\pi$, which deforms $2h\pi$ to 2π continuously. That is, $h(\tau)$ is a continuous function on $\tau \in [0,1]$ with integer values. Since $h(1) = 1$, $h(\tau) \equiv 1$, whence $h = 1$.

This reasoning provides us with the following result:

$$\int_{\partial P} d\theta + \sum_j \alpha_j = 2\pi \tag{3.140}$$

Equations (3.139) and (3.140) obviously yield (3.138), which proves the theorem for polygonal domains situated in the range of a semigeodesic parameterization.

(ii) The second step is to prove (3.138) for an arbitrary polygonal domain P. Then the plane polygon A can be written as a finite union of smaller polygonal domains whose interiors are pairwise disjoint, and the restrictions of the diffeomorphism $F: A \to P$ to these domains provide us with a corresponding decomposition $P = \cup_\rho P_\rho$ (Fig. 3.12). Moreover, because of the continuity of F and the compactness of P, this decomposition can be chosen in such a way that every P_ρ belongs to the range of some semigeodesic parameterization.

It is also important to take the orientations of ∂P_ρ in a coherent manner, that is, such that a common edge receives opposite orientations in the two boundaries to which it belongs. Indeed, this can be done for A, which is a usual plane polygon, and we may assume F to be orientation preserving without any loss of generality.

Let us denote by a_{j_ρ} the vertices of P_ρ, by α_{j_ρ} the corresponding external angles, and by

$$\beta_{j_\rho} = \pi - \alpha_{j_\rho}$$

the corresponding internal angles of P_ρ. Then (3.138) holds for every P_ρ, that is,

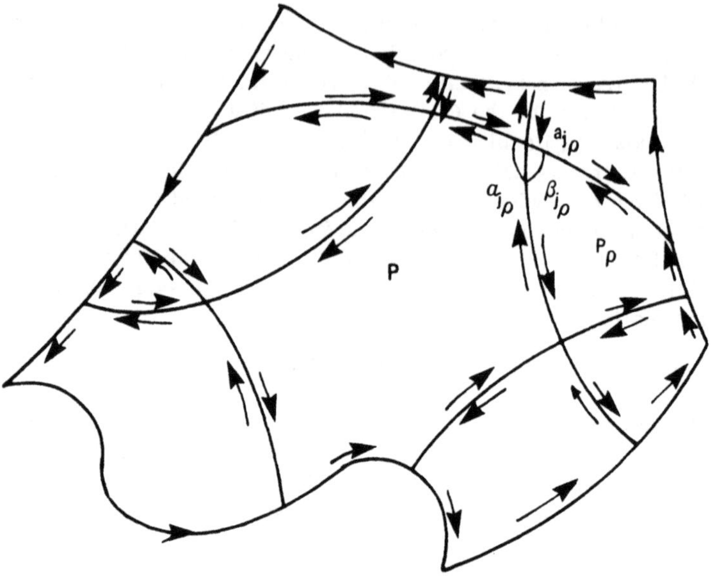

Figure 3.12

$$\iint_{P_\rho} K\, d\sigma = 2\pi + \sum_{j_\rho} (\beta_{j_\rho} - \pi) - \int_{\partial P_\rho} \kappa_g(s)\, ds \tag{3.141}$$

Furthermore, let us denote by f the number of the domains P_ρ, by e the total number of edges of all the P_ρ (a common edge is not counted twice), and by v the total number of vertices (each vertex counted once). A famous Euler formula, known from elementary geometry, states that

$$v - e + f = 1 \tag{3.142}$$

and this is independent of the subdivision of P. (This formula can be explained as follows: If $f = 1$, $v = e + 1$, and the result holds; a subdivision by one new edge increases f by 1, e by 3, and v by 2, and $v - e + f$ remains unchanged. A good induction argument, transforming this explanation into a proof, can be provided.)

Now by taking in (3.141) the sum over all the indices ρ, we get

$$\iint_P K\, d\sigma = \sum_\rho \iint_{P_\rho} K\, d\sigma = 2\pi f + \sum_\rho \sum_{j_\rho} \beta_{j_\rho} - \sum_\rho v_\rho \pi - \int_{\partial P} \kappa_g(s)\, ds \tag{3.143}$$

where v_ρ is the number of the vertices of P_ρ. The last term equals $\sum_\rho \int_{\partial P_\rho} \kappa_g\, ds$ since a common edge of two P_ρ contributes to this sum two terms differing only in sign, which therefore cancel. (This happens because of the opposite orientations of the common edge, as explained above.)

(iii) We have, therefore, to discuss the result (3.143). The vertices of the P_ρ are of three types: vertices that are at the same time vertices of P, vertices that are interior to the edges of P, and vertices that are interior to P. We denote the corresponding angles with one, two, and three primes. Then we have

$$\sum_\rho \sum_{j_\rho} \beta_{j_\rho} = \sum_j \beta'_j + \sum_k \beta''_k + \sum_h \beta'''_h = \sum_j \beta_j + \pi v'' + 2\pi v''' \tag{3.144}$$

where v'' is the number of vertices of the second type and v''' is the number of vertices of the third type, and $\Sigma_j \beta_j$ is the sum of the internal angles of P. We also have

$$\sum_\rho v_\rho \pi = \pi v' + 2\pi v'' + 4\pi v''' \tag{3.145}$$

where v'' and v''' are as above, and v' is the number of the vertices of P.

The sum (3.145) can also be evaluated in another manner: by grouping together the vertices of the same edge. An interior edge (i.e., an edge that is common to two domains P_ρ) has a total contribution of 4π (π for each vertex in each domain P_ρ), whereas an exterior edge contributes only 2π (it belongs only to one domain P_ρ). But in this computation, the contribution of each vertex is counted twice since each vertex belongs to two edges of the same P . Hence

$$\sum_\rho v_\rho \pi = \pi e' + 2\pi e'' = \pi v' + \pi v'' + 2\pi e'' \tag{3.146}$$

where e' is the number of the exterior edges, e'' is the number of the interior edges, and the equality e' = v' + v'' is obvious.

From these results we deduce that

$$\begin{aligned} 2\pi f + \sum_\rho \sum_{j_\rho} \beta_{j_\rho} - \sum_\rho v_\rho \pi &= \Big(\sum_j \beta_j + \pi v'' + 2\pi v'''\Big) - (\pi v' + \pi v'' + 2\pi e'') + 2\pi f \\ &= \sum_j (\beta_j - \pi) + 2\pi f + 2\pi(v''' - e'') \\ &= \sum_j (\beta_j - \pi) + 2\pi f + 2\pi(v''' + v' + v'' - e' - e'') \\ &= \sum_j (\beta_j - \pi) + 2\pi(v - e + f) = 2\pi + \sum_j (\beta_j - \pi) \end{aligned}$$

If we substitute this in (3.143), we obtain

$$\iint_P K\, d\sigma = 2\pi + \sum_j (\beta_j - \pi) - \int_{\partial P} \kappa_g(s)\, ds \tag{3.147}$$

and this is equivalent to (3.138). Q.E.D.

Formula (3.147) can also be considered as the Gauss-Bonnet formula, but expressed by means of the internal angles β_j.

COROLLARY 3.4 If ∂P consists of geodesic arcs (i.e., P is a "geodesic polygon"), the sum of its internal angles is

$$\Sigma \beta_j = (e - 2)\pi + \iint_P K\, d\sigma \tag{3.148}$$

where e is the number of the edges of P. In particular, the sum of the internal angles of a geodesic triangle is

$$\Sigma \beta_j = \pi + \iint_P K\, d\sigma \tag{3.149}$$

For locally flat surfaces, we get the usual $\Sigma \beta_j = \pi$, but this is not true for general surfaces. If K = const, (3.149) yields $\Sigma \beta_j = \pi + KA(P)$, where A(P) is the area of the triangle. We see that, for example, in the Lobatchewsky geometry (see the end of Sec. 3.5), the sum of the angles of a triangle is less than π—an essential feature of that geometry. Contrary to the above, on a sphere the sum of the angles of a triangle defined by arcs of great circles is larger than π.

Now we shall use the Gauss-Bonnet formula to derive some very important information about compact orientable surfaces in E^3. Let us assume that we have a compact orientable surface S, and that it can be decomposed into a finite union of polygonal domains $S = \cup_\rho P_\rho$ with the same properties as the decomposition $P = \cup_\rho P_\rho$ encountered in the proof of Theorem 3.12. This will be called a triangulation of S, and for every such triangulation T we have some numbers v, e, and f of vertices, edges, and domains also called faces), respectively.

DEFINITION 3.20 The number

$$\chi_T(S) = v - e + f \tag{3.150}$$

is called the Euler-Poincaré characteristic of the surface S with respect to the triangulation T.

THEOREM 3.13 (Gauss-Bonnet) The Euler-Poincaré characteristic of a compact orientable surface does not depend on the triangulation T, and it is given by the formula

$$\iint_S K\, d\sigma = 2\pi\chi(S) \tag{3.151}$$

Proof: If we prove (3.151), we have also the independence of χ on T, which explains the notation $\chi(S)$ introduced in (3.151).

To get the formula (3.151) for a given T, we simply apply the Gauss-Bonnet formula (3.147) to each of the P_ρ, add the results, and compute the sum as done in the final step of the proof of Theorem 3.12. As a matter of fact, things are even simpler since every vertex and every edge is now interior. (For example, think of a sphere.) By this procedure, we get

$$\iint_S K\, d\sigma = 2\pi(v - e + f)$$

Q.E.D.

In view of a penetrating topological result, every compact orientable surface S admits triangulations. Hence every such surface has a well-defined Euler-Poincaré characteristic $\chi(S)$, which is given by (3.151).

DEFINITION 3.21 The number $g(S) = 1 - \chi(S)/2$ is called the genus of the surface S.

This allows us to provide a classification of the compact orientable surfaces in accordance with their genus. The following theorem is a complex topological result, which we cannot prove here.

THEOREM 3.14 Two compact connected orientable surfaces of the same genus are diffeomorphic and conversely.

This means that the genus classification is exactly the classification of the compact orientable surfaces up to diffeomorphisms.

Let us give a more intuitive description of the genus. First, if we take the sphere S^2, it can be triangulated by projecting from its center an inscribed tetrahedron. This triangulation has $v = 4$, $e = 6$, and $f = 4$, whence $\chi(S^2) = 2$ and $g(S^2) = 0$. Consequently, the genus zero class is the class of surfaces that are diffeomorphic to a sphere.

Now we construct a new surface by cutting out two holes in a sphere and by gluing up a handle as shown in Fig. 3.13. It is easy to understand that the surface obtained can be deformed to a torus ("a sphere with one hole"), and therefore that it is diffeomorphic to a torus. (Think of the surface as if it were made of some elastic material.) In its turn, the torus can be thought of as the deformation of a box with a hole such as that shown in Fig. 3.14. The natural triangulation of Fig. 3.14 provides a triangulation of the torus. For this triangulation we get (simply by counting in Fig. 3.14)

$$v = 16 \quad e = 32 \quad f = 16$$

which yields $\chi = 0$, $g = 1$. Consequently, the genus 1 class is the class of surfaces that are diffeomorphic to a torus (or to a sphere with one handle).

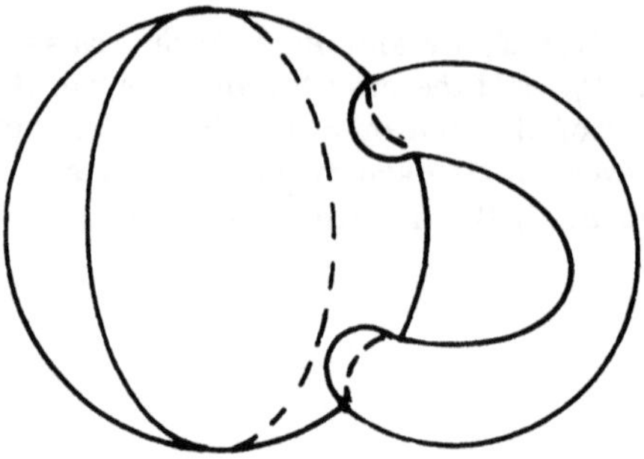

Figure 3.13

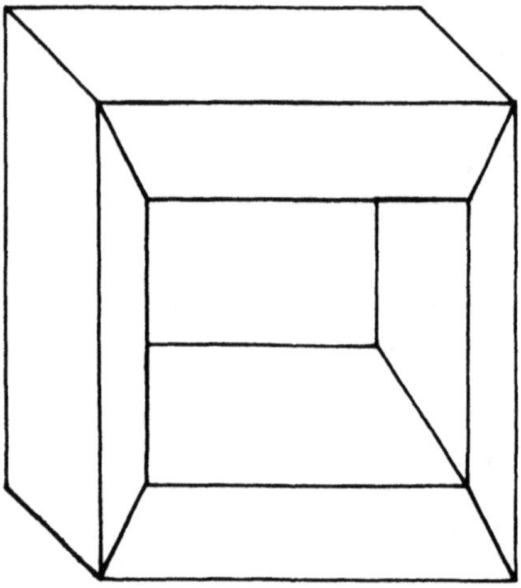

Figure 3.14

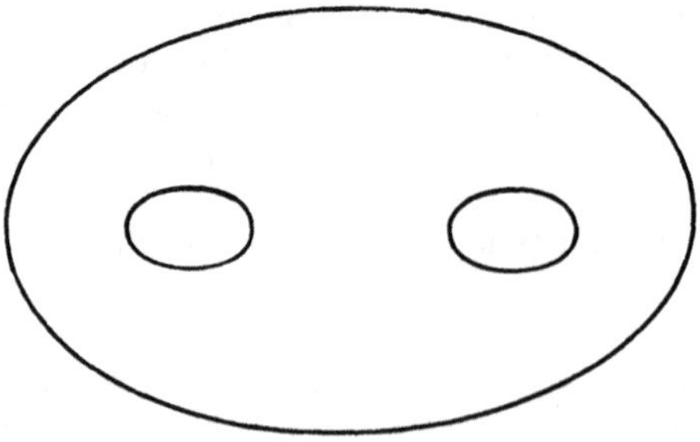

Figure 3.15

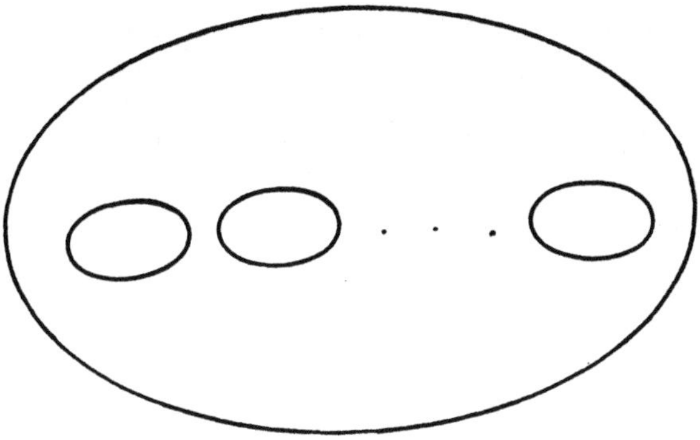

Figure 3.16

Now if we add, similarly, a second handle to S^2, we get something diffeomorphic to Fig. 3.15, called a double torus. To get a triangulation for this it suffices to take two boxes as in Fig. 3.14, cut away the upper face of each, and glue them together along the boundaries of the upper faces. Then two faces, four edges, and four vertices disappear, and χ is diminished by $4 - 4 + 2 = 2$. That is, the characteristic χ of the double torus is -2 and the genus is 2.

We can continue with the procedure in an obvious manner, and it follows that a compact connected orientable surface of genus $g > 0$ is one that is diffeomorphic to a g torus (i.e., a torus with g holes, or, equivalently, a sphere with g handles) as shown in Fig. 3.16. Since there is also another topological result which states that there are no surfaces of negative genus, we have obtained a complete classification of the compact connected orientable surfaces, up to a diffeomorphism.

EXERCISES

3.98 Prove that on a surface S with nonpositive total curvature, there is no polygonal domain P having exactly two sides, where both consist of geodesic arcs.

3.99 Prove that any compact orientable surface of positive total curvature is diffeomorphic to a sphere. [Hint: Prove that it must be a surface of genus zero.]

3.100 Prove that a compact orientable surface that is not homeomorphic to a sphere must have points of all the following types: elliptic, parabolic or planar, and hyperbolic.

3.101 Assume that S is a compact orientable surface of positive total curvature. Let γ_1 and γ_2 be two simple closed geodesics on S. Prove that γ_1 and γ_2 have a nonempty intersection. [Hint: Use Exercise 3.99 and deduce that if $\gamma_1 \cap \gamma_2 = \emptyset$, these two curves bound a region of vanishing Euler-Poincaré characteristic. This contradicts $K > 0$.]

3.102 Prove without the use of a triangulation that the torus has a vanishing Euler-Poincaré characteristic.

References and Suggested Readings

S. S. Chern, Curves and Surfaces in Euclidean Space, in MAA Studies in Mathematics, Vol. 4: Studies in Global Geometry and Analysis, ed. S. Chern. Prentice-Hall, Englewood Cliffs, New Jersey, 1967, pp. 16-56.

M. P. Do Carmo, Differential Geometry of Curves and Surfaces. Prentice-Hall, Englewood Cliffs, New Jersey, 1976.

L. P. Eisenhart, A Treatise on the Differential Geometry of Curves and Surfaces. Dover, New York, 1960.

W. H. Fleming, Functions of Several Variables. Addison-Wesley, Reading, Massachusetts, 1968.

F. E. Hohn, Introduction to Linear Algebra. Macmillan, New York, 1972.

S. T. Hu, Elements of General Topology. Holden-Day, San Francisco, 1964.

H. R. Jacobs, Geometry. W. H. Freeman, San Francisco, 1974.

W. Kaplan, Advanced Calculus. Addison-Wesley, Reading, Massachusetts, 1968.

W. Klingenberg, An Introduction to Differential Geometry. Springer-Verlag, New York, 1977.

R. S. Millman and G. D. Parker, Elements of Differential Geometry. Prentice-Hall, Englewood Cliffs, New Jersey, 1977.

J. Milnor, Topology from the Differentiable Viewpoint. University Press of Virginia, Charlottesville, Virginia, 1965.

M. Stoka and G. G. Vrănceanu, Problems of Differential Geometry. Editura Didactică Pedagogică şi Bucharest, Romania, 1963 (in Romanian).

Index

Milton Keynes UK
Ingram Content Group UK Ltd.
UKHW040054071024
449327UK00019B/550

9 780367 451875